Advanced Techniques in Diagnostic Cellular Pathology

Advanced Techniques in Diagnostic Cellular Pathology

Edited by

Mary Hannon-Fletcher
The University of Ulster, Coleraine, UK

and

Perry Maxwell
Belfast Health & Social Care Trust, Belfast, UK

WILEY-BLACKWELL

A John Wiley & Sons, Ltd., Publication

Wiley-Blackwell is an imprint of John Wiley & Sons, formed by the merger of Wiley's global Scientific, Technical and Medical business with Blackwell Publishing.

Registered office: John Wiley & Sons Ltd, The Atrium, Southern Gate, Chichester, West Sussex, PO19 8SQ, UK

Other Editorial Offices:
9600 Garsington Road, Oxford, OX4 2DQ, UK

111 River Street, Hoboken, NJ 07030-5774, USA

For details of our global editorial offices, for customer services and for information about how to apply for permission to reuse the copyright material in this book please see our website at www.wiley.com/wiley-blackwell

Library of Congress Cataloguing-in-Publication Data

Advanced techniques in diagnostic cellular pathology / edited by Mary Hannon-Fletcher and Perry Maxwell.
p. ; cm.
Includes bibliographical references and index.
ISBN 978-0-470-51597-6
1. Cytodiagnosis. I. Hannon-Fletcher, Mary. II. Maxwell, Perry, 1958-
[DNLM: 1. Cytodiagnosis–methods. 2. Cytodiagnosis–trends. 3. Cytological Techniques–methods. 4. Cytological Techniques–trends. 5. Pathology, Clinical–methods. 6. Pathology, Clinical–trends. QY 95 A2435 2009]
RB43.A38 2009
611'.01815–dc22

2008055188

ISBN: 9780470515976

A catalogue record for this book is available from the British Library.

Set in 11/13 Minion by Thomson Digital, Noida, India.
Printed in Spain by Grafos SA, Barcelona.

First Impression 2009

Cover design adapted from an ISH detection of HER-2 gene amplification in breast cancer.

Contents

List of Contributors

Jim Diamond
Lecturer
Queen's University Belfast
Centre for Cancer Research & Cell Biology
97 Lisburn Road
Belfast, UK

Ian Dimmick
Flow Cytometry Core Facility Manager
Institute of Human Genetics
Bioscience Centre
International Centre for Life
Newcastle Upon Tyne, UK

Mary Hannon-Fletcher
Lecturer
University of Ulster
Coleraine, UK

Merdol Ibrahim
Manager
UKNEQAS ICC & ISH
London, UK

Hilary Magee
Senior Medical Scientist
Department of Cellular Pathology
Adelaide & Meath Hospital, Dublin
incorporating the National Children's Hospital
Tallaght
Dublin, Ireland

Perry Maxwell
Principal Clinical Scientist
Belfast Health & Social Care Trust
Centre for Cancer Research & Cell Biology
Belfast, UK

David McCleary
Doctoral Student
Queen's University Belfast
Centre for Cancer Research & Cell Biology
Belfast, UK

Rachael Natrajan
Molecular Pathology
The Breakthrough Breast Cancer Research Centre
Institute of Cancer Research
London, UK

Anthony O' Grady
Chief Medical Scientist
Department of Pathology
Royal College of Surgeons in Ireland Education & Research Centre
Beaumont Hospital
Dublin, Ireland

John O' Loughlin
Chief Medical Scientist
Department of Cellular Pathology
Adelaide & Meath Hospital, Dublin
incorporating the National Children's Hospital
Tallaght
Dublin, Ireland

Jorge S. Reis-Filho
Molecular Pathology
The Breakthrough Breast Cancer Research Centre
Institute of Cancer Research
London, UK

David S.P. Tan
Molecular Pathology
The Breakthrough Breast Cancer Research Centre
Institute of Cancer Research
London, UK

Preface

Cellular Pathology in recent years has become more closely involved in the direct management of patients with the introduction of molecular technologies and targeted therapies. Through this, we have seen the introduction of specialist pathology. It is the aim of this book to introduce these concepts and the key technologies that are influencing clinical practice today. Throughout this book we show how clinical practice has been affected by these respective technologies and how further development will influence the practice and delivery of Cellular Pathology, which will impact on the patient through targeted therapeutics and diagnostics.

In Virtual Microscopy, we deal with the changing face of the most traditional aspect of Cellular Pathology: that of microscopy itself. For centuries, the glass microscope slide has been the sole method of visualising cells and tissues, but with the addition of computing, imaging and communications technologies, it is now possible to digitise the glass slide and deliver it via the World Wide Web, transforming the ability of the practitioner to deliver a diagnosis, consultation or high-throughput image analysis for biomarker research.

Cytopathology and the introduction of Liquid-based Cytology have greatly improved the quality and thus the sensitivity of the traditional Papnicolaou smear and non-gynaecological cytology specimens. An added bonus of this technology is the surplus of well-preserved material that can be used for additional diagnostic procedures and for research. The introduction of the Human Papilloma Virus vaccination programme may well change the future of the cervical screening programme; only time will tell how effective it will be.

Flow Cytometry has expanded over recent years with the introduction of multicolour cell and protein labelling. Yet it is through understanding the components of the flow cytometer and the properties of the labels that the precise identification of elements important to the practitioner will be enabled.

Immunocytochemistry has seen the continued expansion of the antibody repertoire with diagnostic, prognostic and therapeutic demands. The experience of peer groups and external quality assurance is a vital part in delivering on the promise of identifying patients suitable for targeted therapies.

The search for new biomarkers and therapeutic targets continues and we highlight a strategic means of identifying these key genes and proteins through array Comparative Genomic Hybridization, and outline how this is used in association with companion diagnostic technologies.

In Situ Hybridization has shown promise of being used in the routine laboratory for more than a decade but it is the introduction of the technology for Her2/Neu gene amplification that has realised this potential. Its extension to other biomarkers has followed rapidly.

The content of this book is particularly suitable for students with a basic working knowledge of Cellular Pathology, although each author has given space to providing some basic principles and key references for students less familiar with these new technologies. The book is a suitable text for students to Masters Level in Cellular Pathology but it is hoped that it will support both students and practising scientists who wish to understand more fully the principles and clinical value of the technologies described within.

Mary Hannon-Fletcher, Coleraine
Perry Maxwell, Belfast
October 2008

List of Tables

List of Figures

Supplementary Materials

Supplementary web based material including annotated virtual microscope slides is available with the book. This is provided courtesy of i-Path Diagnostics Ltd and can be accessed online from their website www.pathxl.com.

Access is password protected. To obtain your user name and password please email cellpath@wiley.com

About i-Path Diagnostics: i-Path Diagnostics Ltd is a company specializing in software products for digital pathology, drug discovery and tissue research. The products revolve around whole slide digital scans or virtual slides and provide unique tools and content for a wide range of applications. Media Contact: Eileen Regan +44 28 9032 1110 www.i-path.co.uk

1

Virtual Microscopy

Jim Diamond[1] and David McCleary[2]

[1]*Lecturer, Queen's University Belfast, Centre for Cancer Research & Cell Biology*
[2]*Doctoral Student, Queen's University Belfast, Centre for Cancer Research & Cell Biology*

1.1 Introduction

Virtual microscopy is a relatively new term; however, it has its origins in a much older technology, that of Telemedicine (where Telepathology can be considered to be a subdiscipline). A general definition of this process would be the acquisition, storage and transmission of microscope images from a local site to a remote site for a specific reason. The main reasons for this were usually diagnostic, consultation or educational. The initial concept was probably driven by researchers in disciplines outside those of pathology or microscopy. Many believed that this was an example of a technology looking for an application, and this hindered the widespread acceptance of this technology.

The first telepathology system was developed in the United States in the 1960s when monochrome images were transmitted between locations via a microwave link. However, it was not significantly explored and used until the late 1980s with the advent of readily-available computer equipment. The main problem for telepathology at this time was a general scepticism about its practical use. Pathology is an extremely visual discipline and there were concerns that the technology would be unable to deliver a high enough image resolution for diagnostic accuracy comparable to that being achieved by traditional microscopy diagnosis. However, the rapid growth in high-speed Internet connections, imaging and computing technology has provided a

Advanced Techniques in Diagnostic Cellular Pathology Edited by Mary Hannon–Fletcher and Perry Maxwell

substantial backbone for the provision of a telepathology infrastructure. This has increased the acceptance of this technology among pathologists and it has been shown that it has clear operational and economic benefits to it.

Traditionally there have been two main types of telepathology and it is useful to understand the advantages/disadvantages of these systems within a health care infrastructure. The first (and simplest) type of telepathology is Static Imaging Telepathology, which is also referred to as 'Store-and-Forward' or 'Passive Telepathology'. This is an asynchronous technology in that there is no simultaneous interaction with the microscope slide. In this approach, a pathologist selects images (from a microscope fitted with a digital camera), stores them on a PC and uploads them to other pathologists usually by e-mail. Systems of this type are obviously cheap and easy to implement and may seem useful. However, in consideration there are two disadvantages to this modality: (1) only a limited number of representative images can be transmitted for interpretation. This is useful but it must be remembered that the consulting pathologist is examining a case without any contextual information from the surrounding tissue. There is a concern that a general opinion offered by the remote pathologist may be influenced by the primary pathologist through the selection of representative images; (2) the consulting pathologist (remote) is unable to select the images he/she requires.

The second type of telepathology is Dynamic Telepathology, which is also known as 'Real-time Video Imaging'. During a telepathology session, a microscope is used in conjunction with a PC to send images to a remote computer. This form can be subdivided into two systems: passive-dynamic and active-dynamic. Passive-dynamic systems allow the implementation of real-time pathology across the Internet, where a local pathologist positions and focuses the microscope slide on pertinent regions and a remote pathologist can join the consultation simultaneously. The important point here is that this is an asynchronous communication; the remote pathologist cannot drive the diagnostic session. Active-dynamic systems have the advantage that they are synchronous systems and allow the remote pathologist to control the microscope.

1.2 Digital (virtual) microscopy: equipment for implementation

1.2.1 Automated microscopes

Automated microscopes represent a form of hybrid technology in that they bring together various components within the industry to form a device capable of creating virtual slides without being dedicated to that purpose.

High-specification microscopes are the starting point for this technology and essentially combine imaging (CCD) and computational technology to produce the device. A benefit of this modality is that as the component technologies advance inside the three domains here, it should be relatively easy to integrate them into product advancements. The champion of this technology has been Dr James Bacus, who developed the BLISS system and began working in the area of virtual microscopy in the mid-1990s. The BLISS system (Bacus Laboratories Inc., Chicago IL) is a slide scanner comprising a high-end fully-automated microscope (e.g. Olympus BX61) that rapidly scans glass microscope slides (maximum X60 objective), integrated CCD and a high-specification PC. The system uses a proprietary method of image tiling to transform the glass slide into a virtual slide.

Due to the microscopes used in these systems, they have the capability to produce very high-quality images. Image quality, although important, is however not the only consideration when looking at these machines. The original concept for these was simply to produce an image. In a clinical setting time constraints play an important role, and the ability to batch slides to be digitized is important.

1.2.2 Image production

Image acquisition as in the systems described above is realized via a digital camera. The CCD chip acquires the centre of the field given by the objective, thus reducing optical aberration to a point considered negligible. When the picture is taken, the slide-moving mechanism puts the slide in the next position for image acquisition while refocusing the slide. The whole process takes milliseconds to execute; however, there may be many thousands of iterations.

Image tiling is a technique that requires the capture of multiple small regions of a microscope slide using a traditional CCD camera. Image tiles are subsequently stitched together to create a large contiguous mosaic, or virtual slide, of the entire slide. Due to the sheer number of image tiles required to create the resultant virtual slide, image tiling systems tend to be slow. It can take hours to capture and align the thousands of tiles that are required to create the virtual slide. Consider the example of tiling with a 1000×1000 pixel CCD with a scanning resolution 0.25 μm/pixel. The digitization of a standard coverslip (20×30 mm) area would require the capture of around 9600 image tiles. Assuming that there is no overlap between adjacent tiles, the actual number may be 10 500 + image tiles, to allow for the stitching algorithm to accurately align them. Both BLISS and DotSlide are tiling systems.

Figure 1.1 Olympus DotSlide scanning system

1.2.3 Olympus DotSlide

The Olympus DotSlide system (shown in Figure 1.1) is essentially an automated microscope (Olympus upright BX research microscope) that is computer-driven over the slide to form an image from a Peltier-cooled 1379 × 1032 pixel camera. As with those from other manufacturers, this is a high-throughput machine in that it is provided with a 50-slide loader (five trays of ten slides). This form of slide scanner tends to be slower than the line-scan variety. The manufacturer quotes a figure of less than three minutes per slide with a X20 objective for a sample of tissue measuring 10 × 10 mm. This leads us to consider the question of what a high-throughput machine really is. All manufacturers quote times based on a 15 × 15 mm piece of tissue, quoting times of a reasonable level. However, many tissues that are regularly found in the average slide tray are significantly larger than this. Histology specimens may be this size, but can be 25 × 20 mm, and traditional cytology specimens can occupy the whole cover slip (50 × 25 mm); liquid-based cytology (LBC) specimens, depending on manufacturer, can be less than 25 mm diameter. Under these conditions, times may be significantly higher. According to the Olympus figures this would represent a time of up to 38 minutes to scan a whole coverslip. The problem is exacerbated when a X40 objective is available. This generally allows the scan to be useful if subsequent analysis at the cellular level is required. In the case of Olympus, this would produce a scanning time of up to 150 minutes. Granted this is a worst-case scenario, but it emphasizes the problems in producing a truly 'clinical machine' at present. Many clinical laboratories can produce more than 200 slides per day. All the manufacturers provide high-throughput capability, but the limiting factor here is not how many slides can be batched in the machine but how quickly it can get them processed. My personal experience has been that typically, most average-sized sections take 20–30 minutes per slide to scan and a fully-loaded carousel (e.g. 200 slides) may take 3–4 days to complete, excluding any failures.

1.2.4 Scanners

1.2.4.1 Progressive scan CCD systems

The Nikon COOLSCOPE II is not truly a virtual slide scanner. It should more correctly be referred to as a digital microscope. It has the capability for slide observation and allows digital image capture. It additionally has Internet communications capabilities; this is the major feature that allows the device to act beyond the role of a traditional microscope. Samples can be viewed on networked PCs at remote locations. Additionally, because the essential operations of the COOLSCOPE II can be managed remotely, diagnostic opinions on particular cases can be exchanged within a medical institution or in the global arena.

Mercy Ships is a global Christian-based charity founded in 1978 by Don and Deyon Stephens. Mercy Ships specializes in using hospital ships to provide free world-class health care and community development services to developing nations. Since 1978, Mercy Ships has performed more than 1.7 million services valued in excess of US$670 million and affecting more than 1.9 million people as direct beneficiaries. Volunteers onboard the Mercy Ship *Anastasis* can obtain a second opinion on pathological specimens from an expert located anywhere in the world thanks to satellite technology and the implementation of an onboard telepathology system. Consequently, volunteer staff using a Nikon COOLSCOPE will be able to load samples and subsequently make images available over the Internet. These can be accessed by authorized experts in the United Kingdom, by logging on to a secure dedicated Web page. In its mode of operation the COOLSCOPE allows a pathologist to view a live image. Additionally, it also allows control of the microscope and image acquisition by a local computer. Verbal communication of the diagnosis between the ship's medical team and the remote consultant is made using telecommunications.

1.2.4.2 Zeiss Mirax system

This system was developed in Hungary by a team from Semmelweis University in Budapest and is now marketed under the name 'Mirax' by Zeiss (Figure 1.2). Mirax is currently a family of scanning devices. Mirax

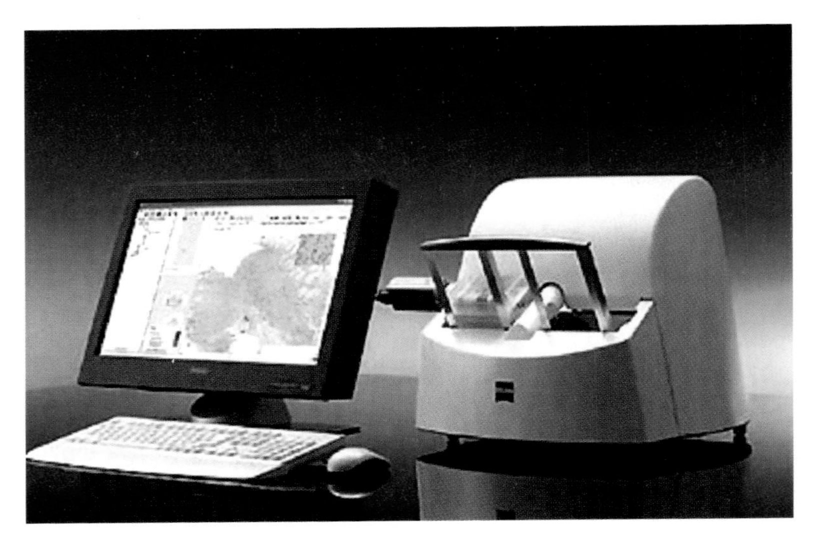

Figure 1.2 Zeiss Mirax scanning system

DESK is a semi-automatic tool for scanning a single slide using a X20 objective. The Mirax MIDI is a research solution and provides the ability to scan 12 slides with a X40 objective. The largest machine in the family is the Mirax SCAN. This is a machine in a similar vein to those from other manufacturers in that it is a fully-automatic system capable of being loaded with 300 slides for scanning up to and including a X40 objective. Fluorescence labelling is currently being used in many research applications. The Mirax SCAN can be upgraded with a fluorescence module, giving it automated fluorescence slide-scanning capabilities. The hardware comprises a 10-position filter wheel and a fibre-coupled fluorescence illumination system.

1.2.4.3 Aperio ScanScope system

Many of the virtual microscopy manufacturers offer varying levels of machine. It is not acceptable to assume that all institutions will have a large demand for slide scanning. The ScanScope GL system from Aperio is an entry-level device allowing small laboratories or university schools to implement a virtual microscopy programme. The machine only offers single-slide scanning on a manual basis. This is suitable for lower-volume environments provided with a X20 objective but with the capability of a final magnification of X400 (via a X2 magnification changer). A similar-specification machine, but offering a higher throughput, is the ScanScope

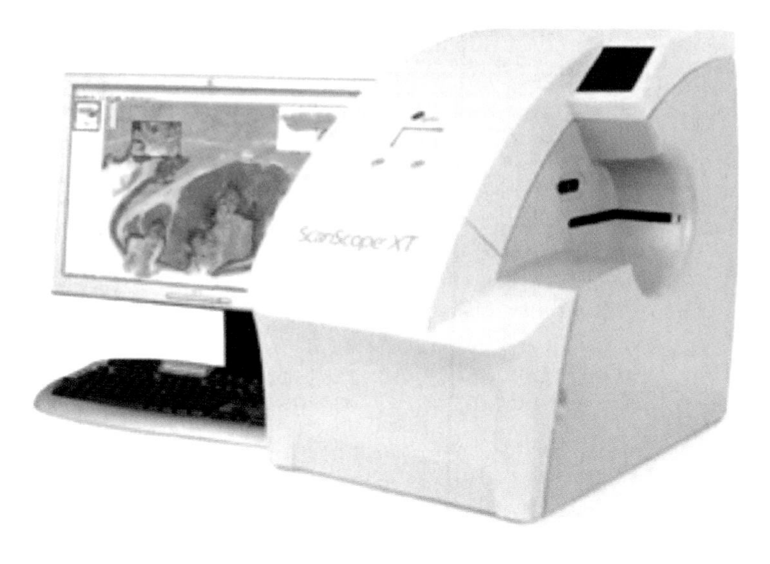

Figure 1.3 Aperio ScanScope scanning system

CS. This machine uses a five-element slide tray, which allows the batch processing of five slides. It claims to be suitable for medium-volume environments, but again, as with the ScanScope GL system, would be only reasonably effective in slide throughput as the slide number grows. The high-end machine from Aperio currently is the ScanScope XT (shown in Figure 1.3). This high-throughput machine offers a slide capacity of 120 slides and is suitable for environments such as hospitals, reference labs, research organizations and pharmaceutical organizations. However, in reality this is probably the machine that is required generally. In my experience the installation of any low-volume scanner simply becomes a victim of its own success. Research/educational groups within organizations tend to come to the scanner with 5–10 slides initially, but rapidly the number of slides increases to 100s as the potential of virtual microscopy is realized. This is not functionally beyond low-volume machines, but it tends to tie up technical support and only five slides can be run overnight. Some specialized pathology preparations, such as blood smears, bone marrow and gram stains, require higher-power scanning using X100 objective (oil immersion), which can be achieved using the ScanScope OS. The scans from the Aperio machine are created in an alternative way to tiling, termed 'line scanning'.

Line scanning accurately moves the slide under a line-scan camera to acquire the image. A line-scan camera is an image-capturing device with a CCD sensor that is formed by a single line of photosensitive elements.

Therefore, unlike area sensors, which generate frames, the image acquisition is made line by line. ScanScope scanners do not use a fixed-area camera to capture thousands of individual image tiles but instead employ linear-array detectors in conjunction with specialized motion-control components, a microscope objective lens and customized optics. As a result, ScanScope scanners efficiently capture a small number of contiguous overlapping image stripes. Whereas image tiling is inherently stop-and-go, ScanScope scanners continuously move microscope slides during the acquisition of imagery data. This ability to capture imagery data while the sample is moving is a key reason why line scanning is ideally suited for rapid slide digitization. Approximately 28 000 image tiles must be captured and aligned to create a seamless digital slide of a 30 × 20 mm area of a slide at a scanning resolution of 0.25 μm/pixel (X40 objective). In contrast, only 60 image stripes must be captured and aligned using the line-scanning method employed by the ScanScope. This example illustrates one of the fundamental advantages of line scanning: the capture and alignment of a small number of image stripes is dramatically more efficient than the capture of thousands of image tiles.

Currently, line scanning seems to offer the advantage of efficient and fast data capture and creation of a virtual slide. Line-scanning systems also benefit from several advantages that optimize image quality: (1) focus of the linear array can be adjusted on each scan line (tiling systems are limited to a single focal plane per tile); (2) the linear array sensor system is one-dimensional; therefore there will be no optical aberrations along the scanning axis (image tiling systems produce a circular optical aberration symmetric about the centre of the tile; (3) the linear array sensor has a complete fill factor (providing full pixel resolution) unlike colour CCD cameras, which lose spatial resolution because of the interpolation of nonadjacent pixels (e.g. using a Bayer mask).

1.2.4.4 Hamamatsu NanoZoomer system

NanoZoomer Digital Pathology (NDP) (shown in Figure 1.4) is a high-throughput slide-scanning system. It is not offered as a low-throughput machine. It has been recognized that most applications will require batch processing, which is an integral part of the system. Up to 210 slides can be loaded for processing. It differs from the Aperio systems in that it offers some additional functionality, while losing the ability to scan at magnification greater than X400. This for now will probably exclude this machine from the haematopathology and microbiology disciplines, where oil-immersion X100 objective scanning is a necessity. This machine does

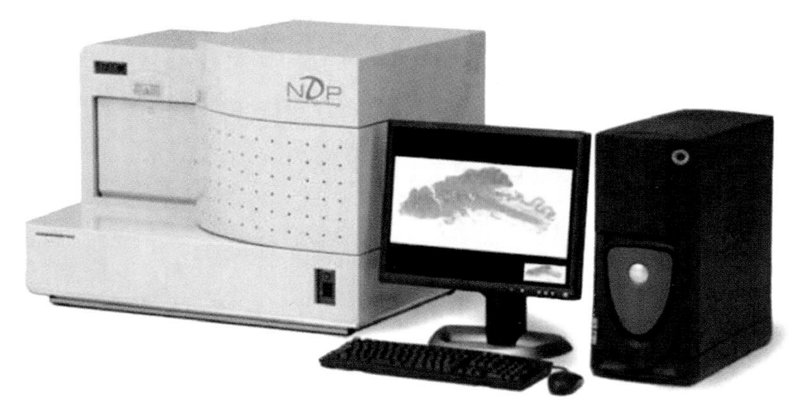

Figure 1.4 Hamamatsu NanoZoomer scanning system

currently have two distinct advantages, namely true 3D *z*-stack scanning and an option to add fluorescence scanning. In addition to standard brightfield applications, therefore, the NanoZoomer system also has fluorescence microscopy capability. Fluorescence microscopy, combined with the NanoZoomer's 3-CCD TDI camera, allows the observation of low-light-level fluorescence tissue samples at high resolution. A realization between all the manufacturers is that virtual microscopy must mimic the traditional microscope exactly, otherwise it will never be truly recognized as a natural development of microscopy. One feature that manufacturers have given minimal attention in the development of their systems is that of focus. There is now an acceptance that all the tissue on a preparation may not nicely fit on one optimized plane of focus (especially appropriate for cytological preparations). They have now recognized this and are implementing 3D scanning (to varying levels). The NanoZoomer is a true 3D scanner (in that it incorporates registered *z*-stack information into the image file) and can scan images in a specified number of slices in the *z*-axis and at a specified distance apart. There is no real limit on the numbers here, however it should be realized that the more slices there are, the larger the already huge image can be made. Typically an LBC preparation slide can be in the order of 600 Mb as a single-plane image; this can be scaled by the number of slices in the 3D image. Typically around 11 (centre + 10 top/bottom) slices are required to gain an acceptable result, although 20+ are probably required to gain any of the diagnostic subtleties observed on the traditional microscope. No real studies are available on how many slices are required; it is simply a trade-off between storage, scan time (this may be many hours) and keeping an optimal degree of accuracy in the image.

Figure 1.5 D-Metrix DX-40 scanning system

1.2.4.5 D-Metrix DX-40 imaging system

The DX-40 slide scanner (shown in Figure 1.5) has been developed by D-Metrix Inc. (Tucson AZ). This company is a spin-out company from the University of Arizona. It was within the Arizona College of Optical Sciences (then the Optical Sciences Center) that Professor Peter Bartels proposed that through miniaturization, tiny microscopes might be produced and combined to form an optical-imaging 'chip', which could serve as a digital-imaging engine for a very rapid virtual-slide scanner. He suggested that such a device could produce the first sub-one-minute slide scanner. At that time this represented a significant reduction in scanning time (X20 objective or better). Additionally, Bartels suggested that it would be possible to extend the field of view of a conventional light microscope from a millimetre (using a X20 objective) to several centimetres. As the field of view of his proposed optical device could be the width of a glass slide, it would be possible to digitally image a whole glass slide with a single pass of the digital-imaging unit. Massive parallel processing of data would further reduce the processing time for a virtual slide.

The DX-40 instrument attempts to implement these principles into a slide scanner. The array technology here replaces the single objective lens with an array of 80 lenses within one instrument. D-Metrix suggests that a single slide can be imaged in one minute and offers a throughput of 40 slides per hour. With such figures, this would represent a 60-fold increase in slide-scanning capacity.

1.3 The virtual slide format

Virtual microscopy, from its inception, has been growing in strength significantly. However, one area in this domain that requires optimization is that of image format. Virtual microscopy currently lacks a globally-accepted image format. In general, manufacturers are providing scanners with an image format that is tailored to their own instrument and its perceived benefits. This is rather short-sighted as it does not acknowledge the requirement that virtual slides be produced not for themselves but rather to integrate into an already present hospital system where standards for image format are already in place.

When the slide has gone through the scanning process the end result is the virtual slide. This is what could be considered an image file; however, because of the file size, all the manufacturers have taken different approaches as to how this is represented and stored on hardware. There are three general frameworks on which to hang the virtual slide: (1) multiple files – usually thousands of JPEGs (or uncompressed files) in one or several folders. Normally each folder corresponds to a different magnification. This is the format of the BLISS system. As seen earlier, this format has the disadvantage that it will need to store thousands of files to hardware. The management of these files in hardware can become difficult as the number of files stored increases; (2) several files with one or multiple resolutions (usually JPEG). This is the method used by Zeiss Mirax scanner; (3) a single compressed JPEG2000 (Aperio) or JPEG (Aperio, Hamamatsu). It is possible to obtain a single file with multi-resolution information. All the information, including the panoramic image or thumbnail and the captures to different resolutions, is stored in a single physical file. Often, the structure of these files is pyramidal and may be TIFF or JPEG2000.

1.3.1 Image format: TIFF

Virtual slides produced by scanning instrumentation are generally recorded in TIFF format with a suffix appropriate to the manufacturer (ScanScope Virtual Slide (.SVS) and Hamamatsu NanoZoomer (.NDPI)). These images are officially compatible with the TIFF standard, although due to their typically large size they are compressed using JPEG2000 protocol, which is not included in the current release of the TIFF library.

1.3.2 Image format: JPEG

The JPEG image-compression standard is currently in worldwide use, and has become the industry standard in photography today. Initially this format was appropriate for the acquisition and efficient storage of images captured from the traditional optical microscope.

The JPEG quality factor (QF) is defined in the range $0 < QF < 100$ and associates a numerical value with the level of compression applied to the resultant images. As the QF decreases from 100, image compression becomes enhanced. There is an inverse relationship with image quality as the resulting image can be significantly reduced in quality, even to the point where the image is unusable. JPEG uses algorithmic encoding to compress images in an 8×8 pixel block. At the highest compression ratios, the 8×8 JPEG blocking artefact occurs, which masks many of the image features. This artefact can also be seen with moderate compression and the virtual slides viewed under higher magnification.

The compression of images introduces another point of contention within the scientific community, and that is the issue of whether images should have compression applied at all. It has been suggested by many manufacturers and researchers that compression of images by the JPEG algorithm should be limited to those intended for visual display purposes only. Virtual slides that are acquired for scientific research with regard to spatial positions, intensities or colour resolutions should never have loss of image information in the process of removing redundant or unnecessary information (lossy compression).

1.3.3 Image format: JPEG2000

A promising solution is JPEG2000, which has potential advantages for use in virtual microscopy, where multi-gigapixel images are the norm. JPEG2000 is a wavelet-based image-compression standard (Joint Photographic Experts

Group committee) developed as an enhancement to the existing JPEG standard. JPEG2000 provides many features that support scalable and interactive access to large-sized images. These include efficient and unified compression architecture, especially at low bit rates; resolution and quality scalability; region-of-interest coding; spatial random access; and effective error-resiliency.

JPEG2000 uses wavelet encoding to compress images in much larger blocks of configurable size. In general, JPEG2000 is a newer and better technique, yielding higher-quality images with higher compression ratios than JPEG. The only disadvantage is that encoding images with JPEG2000 requires significantly more computer processing time. This generally is not good news for the manufacturers as the quest for speed of scanning receives a setback with JPEG2000. Some of the manufacturers provide this format. However, the compression is done 'on the fly' in the hardware. In tests it has been observed that JPEG2000 seems to provide an image quality which is at least as good as JPEG, if not better. The main advantage here is the storage overhead. Typically a virtual slide can be around 2 Gb (JPEG), whereas the same slide compressed to JPEG2000 will be around 650 Mb. The issue of size here really only manifests itself where scanning is being done for archival purposes; in this case the saving can be significant.

1.4 Image serving and viewing

Two applications that are core to the delivery of any virtual microscopy system are the image server and the image viewer. The significant advantage of virtual microscopy is its ability to deliver images remotely. Due to the size of virtual slides this removes the possibility of simply opening the image, as could be achieved in Photoshop, for example. The image must be delivered to the viewer as part of a region-on-demand process. The viewing software will request a spatial region of an image at a specific magnification and the server will access the file and return the appropriate image information as requested. The viewing software will then place the returned image region on-screen in the appropriate position. Image serving/viewing software is provided by all the major virtual microscopy manufacturers. The problem for the user here is that all products only serve out and view the images scanned by that manufacturer. This will need to change and some companies are now starting to develop scanning platform-independent solutions.

Image serving is going to be a central issue for digital microscopy. Pathologists/scientists can be demanding in their acceptance of any new technology and reasons not to use it tend to come easily to their minds. An issue that is central to this is that of speed of delivery of the image over the Internet. If this technology is to be accepted in a clinical setting the use of the computer must provide the same speed of delivery as the microscope (or similar). Internet traffic will always be a problem in the delivery mechanism. However, there are ways around this.

Buffering the image is potentially a method for significantly increasing the speed of delivery. While a virtual slide is being viewed the viewer will be requesting the surrounding areas of the image. When they navigate the slide, the appropriate region will already be in memory. Buffering around the position where the viewer is located is a passive solution to this problem. A more intelligent solution would be smart browsing, where the viewer knows something about the image and buffers relevant portions of it. This could be achieved from an image-processing standpoint; groups are already investigating this methodology. The reason for this thought is that a pathologist/scientist rarely moves in a random manner about a slide – there is always an area of interest. An alternative approach is that of motion prediction; estimating where the viewer will go and buffering the sections ahead.

1.5 Applications of virtual microscopy

1.5.1 General

Quantification of biological and medical analysis is a major concern in standardizing and improving the efficiency and objectivity of the studies. Virtual microscopy in combination with sophisticated image analysis offers a great opportunity.

With virtual slides, the operator can define accurate protocols and run efficient algorithms on whole slides or specified areas. Under the control of a pathologist, this is a powerful tool that makes scoring more accurate and achieves high throughput to increase statistical significance. This is currently of very high interest, for example, in cancer biomarker expression quantification (e.g. HER2/NEU), and is frequently of importance for both patient health and financial considerations. In the research and pharmaceutical industries these methods become increasingly important in proportion to the number of markers, targets and drugs to test. The embedding of source images, analysis parameters and analysis results in a common database from the beginning allows for powerful data handling, including data mining, cross-correlation studies, automatic report generation and so on.

1.5.2 Tissue microarray (TMA)

There are many limiting factors in any pathology or molecular clinical-analysis study of tissues due to: (1) the non-optimization of slide-preparation procedures; (2) the limited availability of diagnostic reagents used in processing; (3) the usually less-than-optimal patient sample size. The technique of using tissue microarray (TMA) was developed to alleviate these issues.

In constructing a TMA, a hollow needle is used to remove tissue cores (usually 0.6 mm but this can vary) from regions of interest in paraffin-embedded tissues (i.e. clinical biopsies). These tissue cores are then inserted in a recipient paraffin block in a spatial array. Sections from this new block can subsequently be cut and mounted on a microscope slide and prepared under the normal processing schedule. TMAs are commonly prepared for tissue immunohistochemistry and fluorescent *in situ* hybridization. TMAs are particularly useful in the analysis of morphology in haematoxylin and eosin (H&E)-stained preparations.

The use of TMA analysis is now becoming standard practice in research pathology laboratories. Automated preparation allows the operator to deposit hundreds of tissue cores from different individuals on the same slide. This is a powerful technique that enhances throughput and leads to greater statistical significance of the results. However, the standard manual analysis of the slide with a conventional microscope is tedious and time-consuming, and in practice the relevance of the method and the timing of the availability of the results are usually limited by a manpower bottleneck. Digital slide-scanning systems that can produce a virtual slide of a TMA slide, used in combination with dedicated TMA analysis software, are significantly increasing the throughput of these processes.

On large TMAs the manual scoring process can be time-consuming and it is easy for the pathologist/scientist to lose their place. Having the TMA stored as a virtual slide negates this. The virtual slide is essentially a series of discrete images representing the TMA. This allows the user to be presented with each core in turn on-screen, thus avoiding missing cores or, more importantly, becoming out of sequence when screening the slide.

Because the slide is in digital form core information can automatically be stored in a database. The advantage of this is that it will allow the subsequent mining of the data for pertinent statistical parameters of the dataset, with the potential to make this data available across the Internet. All data can be linked to each core and can be made available on demand for the efficient and optimal production of reports.

The ultimate use of virtual slides is in the automated analysis and scoring of slides. This is where the major impact of this technology will lie. The throughput quantity is important to laboratories in the pharmaceutical sector, where thousands of slides with hundreds of cores per slide are scored. Here the automated application of algorithms to the virtual slide is of significant importance. This is covered in Section 1.7.1.

TMAs provide a potentially high-throughput platform for the identification of tissue biomarkers in cancer. While the tools for the preparation of TMAs have developed rapidly, the tools for automated TMA analysis are still in their infancy, with pathologists still playing a key role in the subjective visual interpretation of biomarkers.

1.5.3 Clinical examples: tissue-based morphology

1.5.3.1 The identification of tissue type in prostate histopathology

Prostate cancer is set to become the most common cancer in men. Figures show that its incidence has been increasing since 1971. Around 22 000 cases of prostate cancer are diagnosed in the United Kingdom each year. Statistics reveal that deaths from prostate cancer have gradually declined since the early 1990s, but mortality is still high; 9 500 men die from the disease each year. A continuing challenge to the medical community is to develop successful strategies for the treatment and early diagnosis of prostate cancer. It has been suggested that automated machine vision systems would form an element of this overall diagnostic strategy by providing improved accuracy and reproducibility of diagnosis.

The study in [1] was designed to investigate how image analysis on virtual slides as applied to prostate histology sections could be an early marker for abnormal change. Texture analysis and morphological measurements were made from images to allow the quantization of tissue type. Within the study it was assumed that there would be three main types of tissue: (1) stroma – fibro-elastic tissue containing randomly-orientated smooth muscle bundles that act as a framework to support the prostatic architecture; (2) normal tissue – prostatic tissue with increased amounts of smooth muscle, glandular and/or stroma components; (3) prostatic adenocarcinomas (PCa) – histologically diverse and having more than one characteristic composition.

Both texture and morphological characteristics of the scene were used in the classification of tissue. Texture analysis was appropriate for the identification of regions exhibiting greater homogeneity in structure. Consequently, in the present example, texture was used to distinguish between

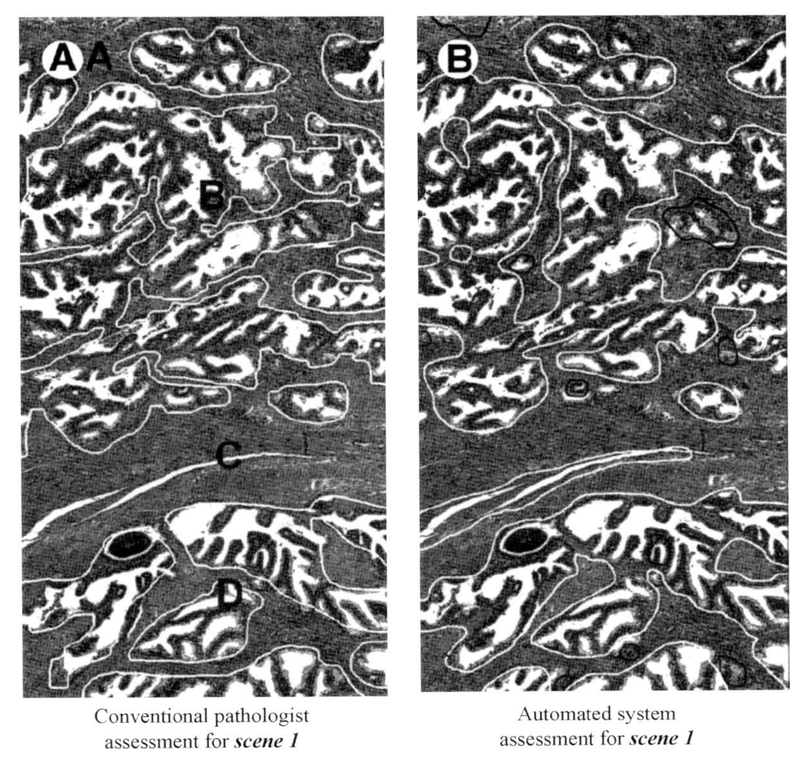

Figure 1.6 Section of prostate histology that has been analysed by both a pathologist and an automated system as showing regions of stroma, benign prostatic hyperplasia (BPH) and prostatic carcinoma (PCa)

stroma and PCa. A morphological approach was applied to the classification of normal tissue, where the glandular tissue is heterogeneous in nature. A pathologist identified representative examples of the three regions defined above and these were used as a training set for the system to classify the tissue. Large regions of tissue were used in the testing of the algorithms (not virtual slides) and some of the results are shown in Figure 1.6. The left-hand side represents the marking of the image (manually) by a pathologist, and the right-hand side represents the output from the system. It can be seen that the region of stroma (A) was identified by the system. The glandular regions (B and D) were identified. Some small regions were misclassified as PCa. The region of stroma (C) was defined, except for misclassification of the urethra region as normal prostatic acinar tissue.

When the algorithmic structure of the system had been tested, the algorithm was applied to virtual slides. The example in Figure 1.7 shows a

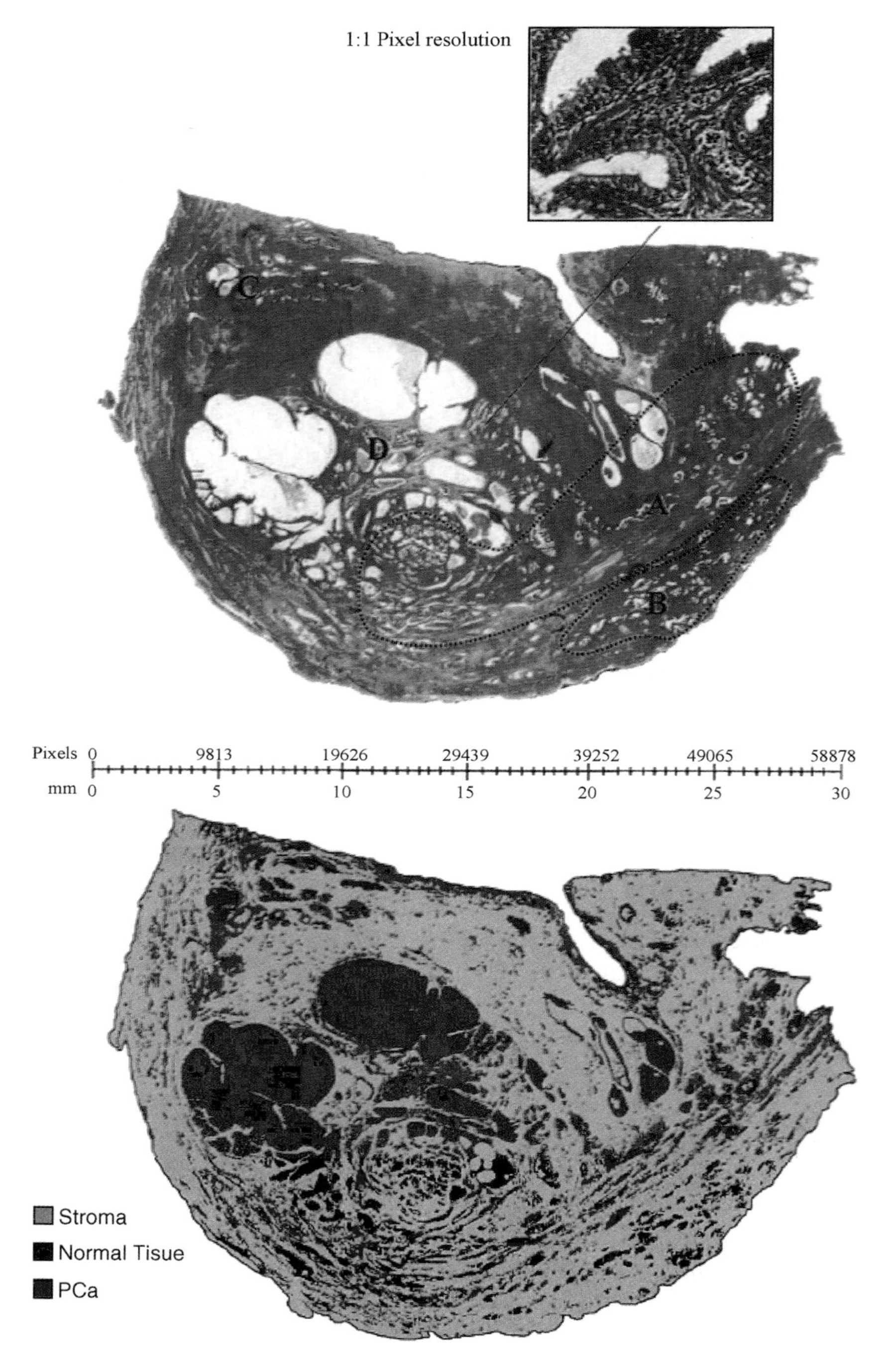

Figure 1.7 Virtual slide showing prostate histology with a computer-generated map of regions exhibiting stroma, benign prostatic hyperplasia (BPH) and prostatic carcinoma (PCa)

piece of prostate histology approximately 30 × 20 mm in size (JPEG format, 58 877 × 42 336 pixels). Regions A and B represent the main areas of interest on this image, with both representing regions of PCa. The PCa in region A has been identified, as has the glandular region D. Region C is an area of lymphocyte aggregate, which has morphology similar to that of poorly differentiated PCa and has been misclassified by the system. Generally the system coped well in the analysis. However, it highlighted some practical constraints on this form of analysis. The process time for this analysis was 5.5 hours, which is clearly impractical for a routine automated system. If such images are to be analysed in this manner then the move to hardware-based systems will be a necessity. Making use of modern high-performance computers, grid computers or field-programmable gate array (FPGA) technology, for example, would result in a substantial speed increase with the move to real-time image processing.

1.5.3.2 The identification and grading of cervical intraepithelial neoplasia (CIN)

Cervical intraepithelial neoplasia (CIN; described in more detail in Chapter 2), also known as cervical dysplasia, is defined as the abnormal growth of potentially pre-cancerous cells in the cervical epithelium. Dysplasia is used to describe histological tissue changes and dyskaryosis is used to describe cellular changes. Most cases of CIN remain stable and change is held within the epithelium, or is eliminated by the immune system without intervention. However, a small percentage of cases progress to become invasive cervical cancer, usually cervical squamous cell carcinoma (SCC). Figure 1.8 shows a section of the cervix that is stained with H&E.

The grading of CIN is problematic, with poor inter/intra-observer reproducibility. CIN represents a morphological continuum and biopsies displaying CIN are classified into three grades (Chapter 2, Figure 2.9). There are also difficulties in reliably distinguishing low-grade CIN from its reactive stimulants such as koilocytosis (Chapter 2, Figure 2.7). CIN 1 represents low-grade dysplasia and is confined to the basal 1/3 of the epithelium. CIN 2 is moderate-grade dysplasia confined within the basal 2/3 of the epithelium. CIN 3 represents high-grade dysplasia and occupies >2/3 of the epithelium, and may involve full thickness (also referred to as carcinoma *in situ*). CIN 3 is a precursor to invasive carcinoma of the cervix.

Due to the size of virtual slides and the time constraint in processing them it is important to optimize the processing conditions. Processing must mimic what a pathologist will do in the routine traditional examination of a slide. This can be highlighted in the identification of the

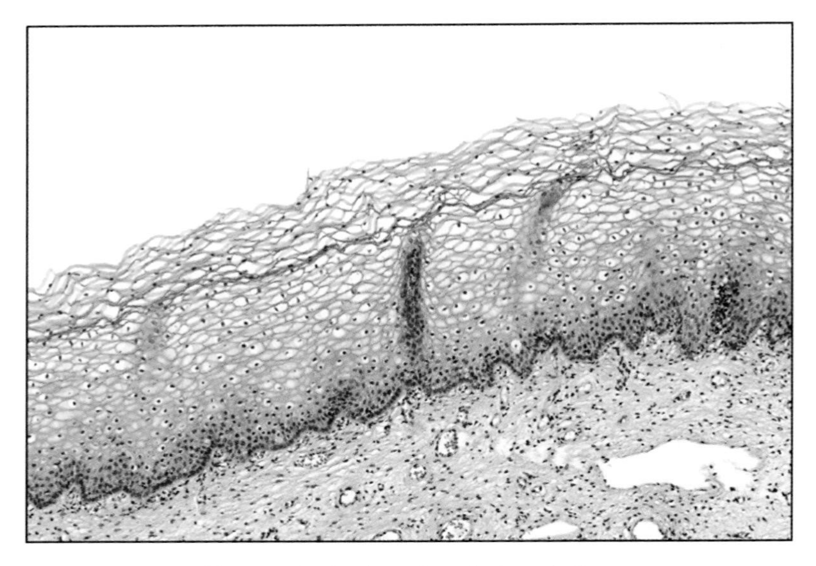

Figure 1.8 Region of cervical epithelium

epithelium, which would be carried out at low-power magnification. On the virtual slide this would be represented by doing any analysis on a subsampled representation of the original image. This is the first step in the image-processing process and allows us to partition the image, in that all processing only needs to be initiated on tissue within the epithelial limits. Figure 1.8 represents a section of the epithelium.

Twenty H&E-stained cervical histological slides were selected for this study [2]. The selected cases were examined and annotated by an experienced pathologist as highlighting normal epithelium, koilocytosis, CIN 1, CIN 2 and CIN 3. All these glass slides were scanned with a X40 objective using a ScanScope CS scanner (Aperio CA) and archived in 24-bit colour JPEG format. A X2 magnification was found to be optimal for use in the identification of the epithelium. This represented a reduction of the image data size to 0.25% of the original.

Identification of the epithelium was achieved using image texture analysis. Image texture, defined as the quantization of the spatial variation in pixel intensities per channel RGB, is useful in many applications and has been a subject of many researchers, particularly in diagnostic imaging. One immediate application of image texture is the recognition of image regions (epithelium) using texture properties.

Prior to evaluating texture features, it is necessary to discover what the main components are inside a typical cervical virtual slide: background; stroma; squamous epithelium, columnar epithelium and red blood cells (RBCs).

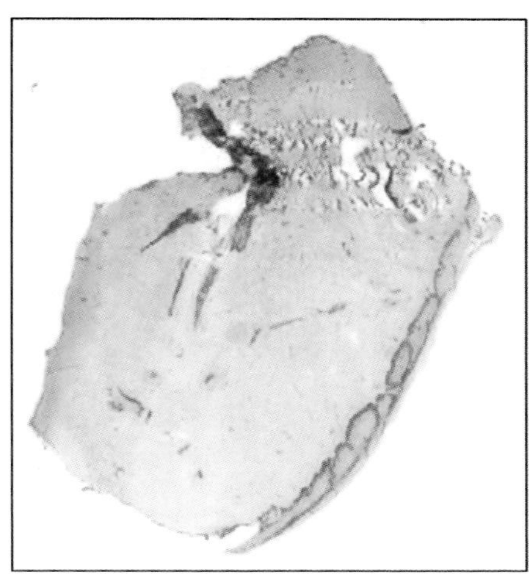

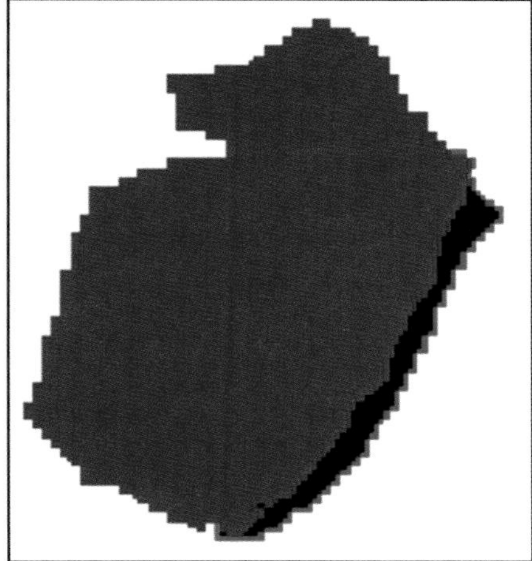

Figure 1.9 Cervical histology with a corresponding computer map highlighting the region of epithelium. Additionally, the basal (blue) layer and surface (green) are indicated

A two-category classification of background vs. tissue component was performed, which requests the least number of texture features and gets rid of all the background, the majority component of a virtual slide. Thereafter, another two-category classification, squamous epithelium vs. stroma, was performed inside the segmented tissue region to leave only squamous epithelium with some parts of misclassified columnar epithelium and RBCs. A post-processing procedure was then applied for the removal of columnar epithelium and RBCs.

Figure 1.9 highlights the identification of the epithelium. The red region is defined as a general area of stroma. The region of the epithelium is highlighted in black. Additionally, the algorithm has determined the orientation of the epithelium. The blue line locates the basal layer and the green line locates the outermost region of the squamous epithelium. The determination of orientation is important as in the next stage of grading we will see how this was necessary to classify CIN.

As we shall see in Chapter 2, the existences and degrees of dysplasia/dyskaryosis associated with CIN are related to the nuclear size (and variation in nuclear size and shape, pleomorphism), nuclear density and texture, and the spatial inter-relationships between nuclei. Therefore, for the grading of CIN it must be prefaced by the segmentation of the epithelial nuclei (Figure 1.10). Having identified the epithelial nuclei, their spatial position within the epithelium and degree of dysplasia must be quantified. This is achieved using a process of Delaunay triangulation and nuclear texture analysis. Having achieved this, we are now able to define the grade of the

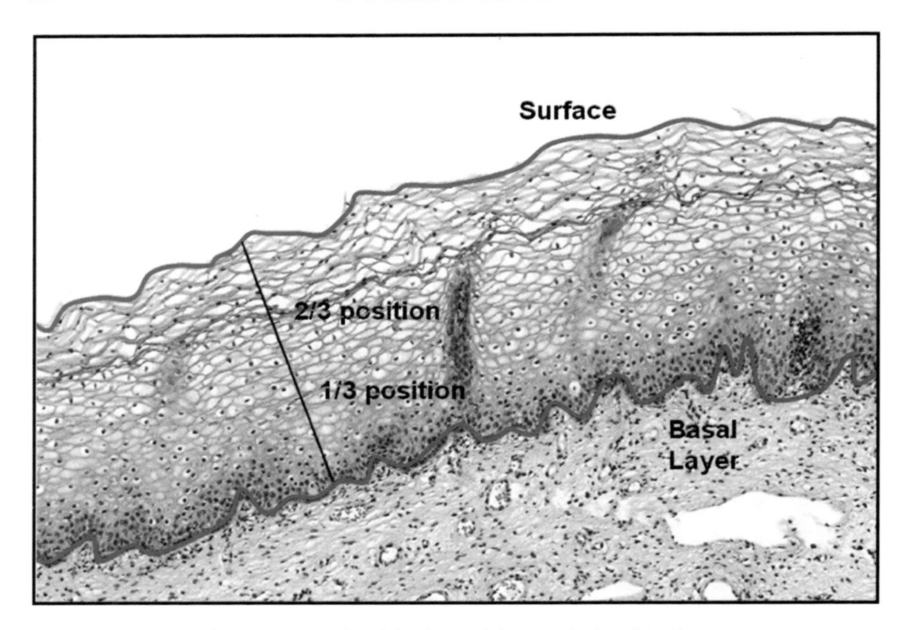

Figure 1.10 Partitioning of the cervical epithelium

lesion in a similar fashion to a pathologist, as to the region of the epithelium occupied by dysplastic cells (i.e. CIN 1 bottom 1/3 occupied, CIN 2 bottom 2/3 occupied, CIN 3 full-thickness occupation). An important point must be made here. A cytology report of mild or moderate dyskaryosis does not always correlate well with the presence of CIN 1 and CIN 2. A deciding factor is often the quantity of CIN rather than the quality. Histological samples may display CIN 3 changes after a mild or moderate dyskaryotic cytology smear result. However, the degree of disparity is not usually so great.

A limiting factor in the exploration of imaging techniques for diagnostic pathology is the number of dimensions of the image that can be considered. All studies generally show 'proof of concept' with small images captured from a traditional CCD camera. Recently this has changed with the introduction of virtual microscopy. This innovative technology now makes it possible for researchers in imaging and pathology to take a significant step towards making their efforts diagnostic and not simply proving that measurement can be made. Currently the use of virtual microscopy in image analysis is making inroads into the arena of diagnostic pathology. Installations in the United Kingdom are minimal at the time of writing. Consequently this represents a prime opportunity to engage in early innovative research in this field. The majority of installations of virtual microscopy are evaluating it from a visualization standpoint (can it replace the microscope?); however, there is significant scope for downstream research exploring how we can take these images into the diagnostic arena through the development of automated devices for quantitative assessment. A major consideration in implementing such systems will be the computational challenge in processing these images. The current PC/software framework will simply not provide enough computational power, so hardware/high-performance computing implementations will need to be evaluated.

1.6 Virtual microscopy in education

Pathology education for undergraduate medical and biomedical students has traditionally been implemented using microscopy with standard glass slides. Virtual microscopy is now offering an alternative approach to this regime. Both the educator and the student have their views on the effectiveness of this approach.

Virtual microscopy allows viewing of virtual slides by large numbers of students or trainees over a computer network, thus avoiding the necessity for them to be at a particular venue at a set time to attend a teaching or training session. This can lead to a large reduction in the time and expense required to organize and run these sessions. Traditionally, the students and trainees would have viewed images generated by a digital camera mounted on a microscope. Each person would have only been able to view the part of the slide and objective magnification selected by the microscope operator. Using Web-server software it is now possible for each person to view a portion of the virtual slide selected by them at the magnification they choose. Not only are there significant cost savings to be made but the quality of the learning experience is enhanced. With the advent of wireless networks it is possible to extend the accessibility of teaching and training materials to make them available on portable and pocket computers/devices to facilitate 'anytime, anywhere' learning.

Virtual microscopy provides several advantages over the traditional approach: (1) the first, obvious one is that we do not need a microscope. Microscopes are expensive to buy (and maintain) and are specific to a microscopy class. Universities usually have an IT infrastructure in place and the education sessions may now be implemented in general-purpose IT suites; (2) digitized images may be archived to an image repository and served out, facilitating sharing by many students (everyone may use and be examined on the same material). Rare, valuable and one-off specimens such as reference tissue biopsies, which cannot be duplicated or reproduced because of limited samples, can be used for education; (3) There will generally be no breakages as there will be minimal removal of slides from archived storage; (4) students tend to assume that the microscope can simply be turned on and the slide viewed and tend to have little (or any) knowledge on how the microscope should be optimized for use (e.g. Köhler illumination). Virtual slides can always be presented in optimal state for viewing.

Virtual microscopy implementations in education are held currently within the domains of computer-aided learning (CAL) and self-directed learning (SDL). Interest in the use of SDL has increased in recent years and has received significant attention from academics. In many UK universities many new programmes, practices and resources for implementing SDL have been initiated. It is beyond the scope of this chapter to consider the pros and cons of SDL, and valid arguments apply to both camps. However, it is in this domain that virtual microscopy in education finds itself. Many things are known about SDL, but most significant is that it empowers learners to take increasingly more responsibility for various decisions associated with the learning process and to explore this process themselves, usually under the guidance of some computer-based (or approved Web delivery system) mechanism. SDL is becoming an essential element in medical/biomedical education due to the expansion of knowledge, accessibility to information and greater emphasis on reflecting on issues learned. So what is the role for virtual microscopy in this domain? Here virtual microscopy is simply an enabling technology. How we use Internet and computational methodologies as an adjunct to virtual microscopy provides downstream functionality.

1.6.1 PathXL

PathXL is an online platform for hosting and managing virtual microscopy and digital slide media for diagnostic pathology, laboratory medicine and tissue-based research. It is a content-management system that permits the hosting and management of virtual slides. This tool has a wide range of functions that support e-learning, quality assurance and teleconsultation in pathology.

The architecture of PathXL relies on Internet servers providing the information that is presented on the PathXL Web site. Managers within this system have full control over the look and feel of that Web site, how the slide-based content is organized and how it is presented. Additionally, there is no restriction to only managing virtual slides; other content such as documents, standard images, Web sites and so on can be added to the PathXL platform.

PathXL provides a dedicated online viewer for virtual slides. This viewer can set interface notes and annotations, record movement through the

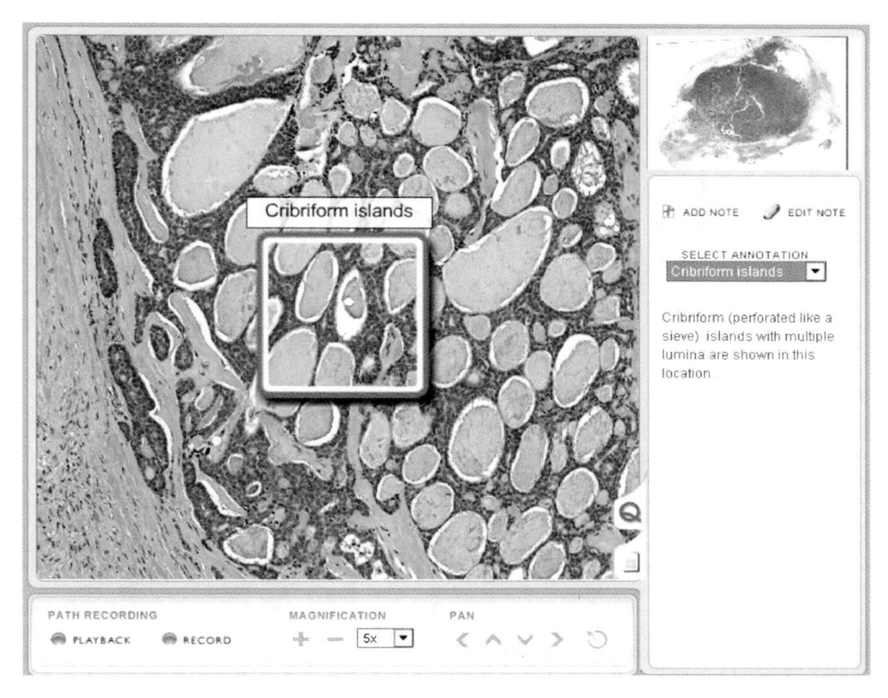

Figure 1.11 Example of an annotation within the PathXL platform

slide, set questions on slides or slide sets, and store answers to a central database for later review. Administration and configuration of PathXL occurs at two levels: (1) administration backend allows the operator to construct modules, add cases, modify content and control user privileges, in addition to server configuration, account management and access to database results; (2) the PathXL viewer provides the interface for setting annotations and questions on the slide, as can be seen in Figures 1.11 and 1.12. The system has been designed to make sophisticated administration of the PathXL account as easy as possible. So while there is control of the settings, this complexity does not interfere with the easy set-up and management of the content. PathXL is currently being used for undergraduate and postgraduate education and training, slide seminars, quality-assurance programs, tissue-based research, TMA management and tissue archiving.

1.6.2 Queen's University Belfast perspective

Queen's University Belfast has for over one hundred years taught microscopic anatomy in a traditional way, using microscopes to view glass slides.

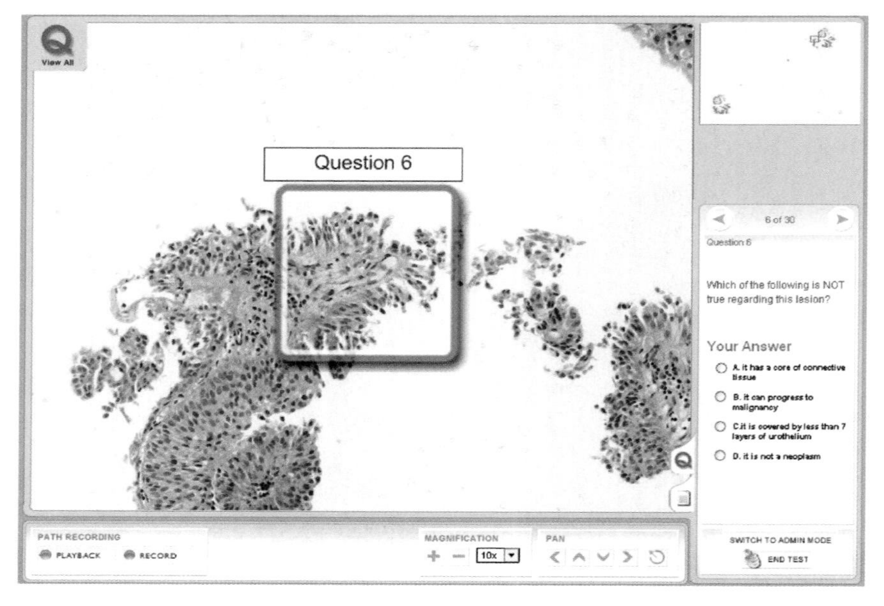

Figure 1.12 Example of question setting within the PathXL platform

2006 saw the introduction of virtual microscope slides in place of glass slides and microscopes in micro-anatomy practical classes, with the installation of two ScanScope CS scanners and one NanoZoomer scanner (Hamamatsu, UK). Virtual slides can also be viewed directly online using a standard browser, providing remote access to valuable teaching material via the PathXL platform. PathXL was used in the current study to host, manage, author and deliver virtual slides to the student population. Here a dedicated Web viewer interacts with an image server, which serves out appropriate regions of the slide image. In this way users can view the whole slide in real time without the need to download the entire image. The management and authoring software is entirely Web-based, allowing virtual-slide-based educational modules to be developed anywhere online. The virtual slides (scanned with a X40 objective) can be stored on an independent server, which can be remote from the authoring software. The front-end virtual-slide viewer is written in Flash (Adobe Systems Inc., San Jose CA) and does not require any specific downloads to the client machine. In this way, students can access the educational slide online within a classroom setting or anytime, anyplace from their home computer. Since the viewing software is browser- and platform-independent, students can view slides online on both Windows PCs and Macs. Figure 1.13 highlights an education session delivered via the PathXL platform.

Figure 1.13 PathXL being used in an educational setting

There are many reasons for implementing virtual microscopy, as mentioned above, but an additional (and fundamental educationalist) one is to encourage greater engagement with microscopic anatomy by students. Although traditional microscopy is an important foundation subject that supports the understanding of key areas such as physiology, gross anatomy and pathology, many students perceive it as dull and it has been difficult to motivate them sufficiently to participate actively in practical classes.

When virtual slides were first introduced into classes, we surveyed the opinions of second-year medical students, who by that stage had had a year studying using conventional microscopes. Out of 136 students who replied, 88% said they would prefer to use virtual microscopy, although 11% were of the opinion that implementing virtual microscopy would lead to a loss of microscopy skills which they believed to be important. The majority of students (99%) found the virtual slides easy to navigate and 93% thought that virtual image quality was at least as good as that of a traditional microscope. More importantly, and a point of significance, is that tutors reported that students showed more interest in the slides and hence the whole anatomy course. As a result it has been possible to provide modules specifically tailored to the needs of students on different degree pathways (e.g. medicine, dentistry, biomedical science and nursing) in a modern and optimal environment that stimulates learning.

1.6.3 Training

InView is a new software platform aimed at training pathologists and biomedical scientists in diagnostic cytology and histology. InView uses a completely innovative approach to education in diagnostic practice through case-based diagnostic simulation using virtual microscopy, decision analysis and instructional video. InView is developed by i-Path Diagnostics Ltd., Belfast, UK. All educational content is approved by The Royal College of Pathologists Australasia (RCPA) and covers many diagnostic scenarios (breast cytopathology Pap & Giemsa, urine cytopathology, cervical histopathology, breast histopathology, skin histopathology, prostate histopathology, salivary gland histopathology and neuro histopathology).

Products such as InView are emerging as a consequence of the virtual microscopy revolution. Before the advent of virtual microscopy, educational products such as InView would have been impossible. InView and products like it rely on the distribution (or Web delivery) of content. In this case the bulk of the content is the virtual slide. Before virtual microscopy, physical glass slides would have needed to be delivered with each instance of the application, clearly a factor against the delivery of any such tools.

InView as a tool has grown out of medical decision support research, where investigations were made into how medical/scientific professionals actually make diagnostic decisions. The ability to quantitatively define how decisions are made provides an inroad into being able to develop computational methods to assist in the process of diagnosis.

InView was developed around a methodology derived from the artificial intelligence community – Bayesian belief probability and its extension into belief networks. The explanation of how these modalities work is beyond the scope of this chapter but suffice it to say InView provides a well-recognized methodology for mimicking the diagnostic process numerically. The only real problem with decision support within the medical community is that of 'do they really want it?' The answer is a vague no. Handing over responsibility to a machine to make a diagnostic decision is going just a bit too far. Public acceptance of machines helping in the diagnostic process is at about the same level; society will always demand a personal decision on diagnostic matters. So where does this leave decision support? The methodology of InView provides a framework for the establishment of a training regime [3].

InView provides an interface which implements diagnostic simulation and guides the user through the diagnostic process with the aim of showing the user how to make diagnostic decisions. A pathological decision in any diagnostic scenario is usually made by observing a sequence of clues and observing the severity of each clue in the case. Here subjectivity comes into

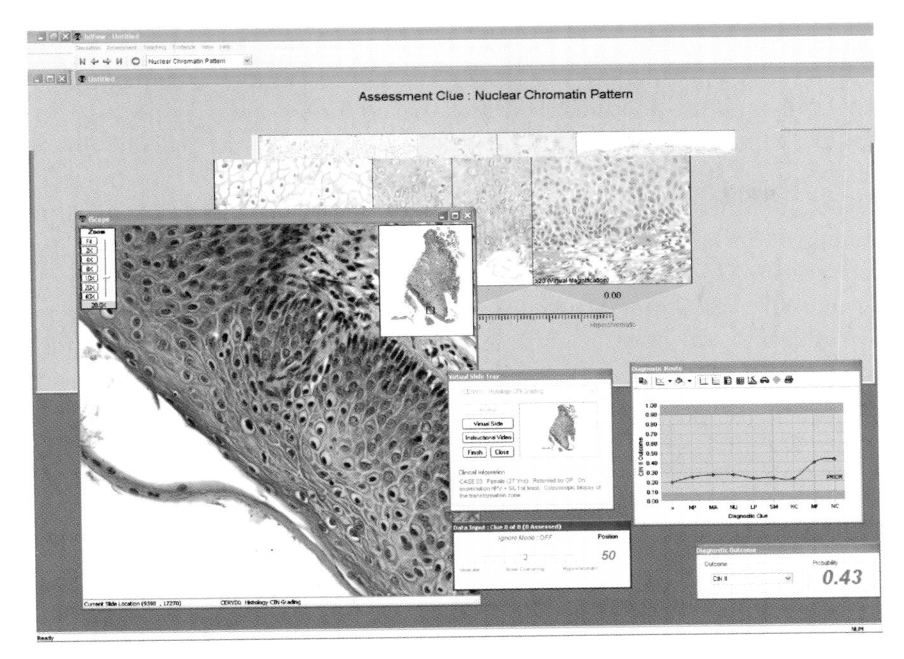

Figure 1.14 InView interface

the equation; this is where individuals are unsure and make errors of misinterpretation or omission. InView attempts to teach the user to omit nothing and provides reference images to show them how to interpret varying clues. At each point, the user is asked to grade the clue presented, and the system responds with the diagnostic outcome based on that evidence. The interface to InView is shown in Figure 1.14, where two reference images can be seen, together with the virtual slide of the case in question.

The number at the bottom right (0.43) is the final belief that the case falls into a particular diagnostic category after the assessment of all clues. More important is the graph above, which shows that the user has arrived at the correct answer for the appropriate reasons. This is achieved by comparing the graph produced against a reference (correct) graph for that case.

InView is integral to the PathXL platform and highlights that virtual microscopy is only a stepping stone in the implementation of downstream applications, from a starting point of simply having a slide.

1.6.4 Proficiency testing/certification

Pathology, whether practiced by the pathologist or by the biomedical scientist, is generally a well-defined specialty and could be considered to be implemented in a similar manner by all practitioners, independent of country. However, there are major differences in the training that is required and the manner in which practitioners are certified. In the United States, 85% of practitioners are certified, and usually this is considered a measure of quality. Certification is through the American Board of Pathology (ABP), which has been operating since 1936. A recent change in the board's approach is the implementation of maintenance-of-certification (MOC), and over the coming years there will almost certainly be an implementation of such programmes worldwide. These programmes will address many issues, such as learning, but an important consideration will be the monitoring of cognitive expertise, which essentially means the maintenance of diagnostics skills. This, as on most programmes, is currently implemented via examination at a remote site on an annual basis, although there is an acknowledgement that this could be moved to a World Wide Web modality given security considerations. Virtual microscopy may have an important role to play in this form of monitoring. The ABP has initiated the introduction of virtual microscopy into its programmes. Currently a small percentage of questions are offered virtually, usually for biopsies where there would not be enough material to produce sufficient glass slides. However, this is likely to snowball and the likelihood is that before long the complete cycle of certification and testing from any organizational body will be online with the use of virtual microscopy.

1.7 Computational aspects of virtual microscopy

Image analysis is a powerful technique used to extract useful quantitative information from images; this usually refers to the analysis of digital images using computerized techniques such as pattern recognition (morphology), texture analysis, densitometry and digital signal processing. Since medical images are composed of distinct shapes, patterns and colours, image-processing techniques can be used to quantifiably analyse them. By merging image analysis and virtual microscopy we can unlock the true potential of virtual slides. However, image analysis of virtual slides is not a straightforward task. Unlike radiology images, virtual slides cannot be opened by imaging software such as general-purpose manipulation tools (Adobe Photoshop). Some platforms are beginning to come into use that can deal with virtual slides (ImagePro Plus). This is due to two factors: image size and format.

Virtual slides can be many gigabytes in size and contain more than a billion pixels, and as a result automated image analysis is extremely time-consuming. As we have seen from the example of the identification of tissue type in prostate histopathology above, automated analysis can take up to 5.5 hours, obviously making routine automated analysis of an entire slide impractical [1].

Another factor which makes image analysis of virtual slides challenging is image format. As we have discussed above, virtual slides are generally compressed using JPEG/JPEG2000, with each scanner manufacturer using its own proprietary image file format, for example Hamamatsu virtual slides are. NDPI (NanoZoomer Digital Pathology) files, which are loosely based on stripped tiff images, typically JPEG-compressed and use the blue, green, red colour model. Aperio store their images as.SVS (ScanScope Virtual Slide) files, which are standard pyramid TIFF images, JPEG- or JPEG2000-compressed and use the red, green, blue, alpha colour model. A detailed understanding of this underlying file format, compression type and colour model is necessary before any image analysis can be conducted. With an increasing number of scanner manufacturers, each with a unique image format, algorithm development is both difficult and time-consuming. Ideally, virtual slides should be in a standard open format, promoting and encouraging the development of imaging algorithms.

These problems can be solved in a number of different ways. Aperio has developed a plug-in to allow MATLAB and ImagePro Plus to run algorithms directly on a ScanScope Virtual Slide image. However, algorithms developed using this approach won't run on any slides scanned using any of the other scanning manufacturers and obviously this approach will not solve the speed issue. The only way to rapidly analyse slides is to use dedicated high-performance hardware. There are a wide variety of technologies that offer solutions to the problem on the market, which can be grouped together under the following headings: (1) high-performance computing; (2) grid computing; (3) other hardware methodologies.

1.7.1 High-performance computing

A high-performance computer is built to be state-of-the-art in terms of technology, particularly in regards to processing capacity and speed of processing. They have been around since the 1960s and were designed primarily by Seymour Gray. Early supercomputers were simply computers with very fast scalar processors. Today supercomputers are parallel designs based on off-the-shelf processors combined with specialist interconnects and are mainly highly-tuned high-performance clusters (HPC), which can be seen in Figure 1.15.

A high-performance computer is a number of linked nodes (computers) interconnected via a dedicated high-speed network to form a single

Figure 1.15 Example of a high-performance cluster

computer, and is typically more cost-effective than comparable standalone computers. The current fastest supercomputer in the world is the IBM Roadrunner. It is a cluster consisting of 122 400 processor cores, occupies 6000 square feet of floor space and was built for the American Department of Energy to simulate how nuclear materials age. Clusters have performance gains over stand-alone desktop computers since they are parallel. A traditional one-processor core machine can only execute one task at a time, whereas a cluster can execute as many tasks as it has processor cores.

One area where there is an urgent need for rapid automated image analysis is that of virtual slides imaged from TMAs, described in Section 1.5.2. As we have seen, TMAs are potentially a high-throughput technology for biomarker discovery and cancer therapeutics. They are a powerful technology allowing the efficient and economic assessment of biomarkers across a large set of tissue samples. However, scoring of TMAs is by visual interpretation, representing a significant bottleneck in the process of these resources. Manual scoring of TMAs is also error-prone and subjective. These problems can be addressed by a high-performance automated system. By combining virtual microscopy, TMAs and high-performance clusters, a system for the rapid, high-throughput automated analysis of TMAs can be developed.

The biomedical imaging and informatics research group at the Centre for Cancer Research & Cell Biology (CCRCB), Queen's University Belfast has been developing such a system for the rapid analysis of TMAs using a high-performance cluster. The system works by identifying each of the cores in the TMA using a semi-automated core-identification system. The coordinate positions of the cores on the TMA virtual slide are then stored in a database. Now that the position of each of the cores is known, the high-performance cluster can begin the image analysis. The cluster works on the standard manager/worker model, where the cluster is divided up into a single manager and many worker nodes. The manager node is responsible for giving the worker nodes work. The manager node has access to the coordinate position of each of the cores. The worker nodes request work from the manager node, which sends out the coordinate positions of the cores that need to be processed. The worker node then dips into the virtual slide and extracts the next core that needs to be processed, performs the image analysis and sends the result back to the manager node, which stores the results in a database. This cycle continues until all the cores in the TMA have been processed.

1.7.2 Grid computing

Grid computing is another style of parallel computing whereby a virtual supercomputer is formed from a cluster of networked computers acting in unison to perform very large tasks. What distinguishes it from cluster computing is that grids tend to be heterogeneous and geographically dispersed, and to consist of standard desktop computers connected via a standard office Ethernet or through the Internet. It is most commonly used for computationally-intensive scientific tasks such as drug discovery, economic forecasting and the search for extraterrestrial intelligence.

Grids work by dividing a task into subtasks, which are farmed out to spare processors. The progress of the subtasks is monitored and if they fail the task is restarted. Results are collected and collated by the grid computing engine. The Oxford Cancer Project [4] used grid computing for computational drug discovery. A specialist screensaver was developed to harness the computing power of millions of computers to build a database of 3.5 billion molecules with known routes to synthesis. These compounds were screened using a process known as 'virtual screening', where analysis software identified molecules that interacted with proteins and determined which of the molecular candidates was likely to be developed into a drug.

IBM is very influential in the area of grid computing and has established a world community grid to solve problems that benefit humanity. The grid works by utilizing thousands of PCs throughout the world. Idle computers request data from a specific project on the world community grid server. They perform some computations on this data and feed the results back into the community server. One project that is being run on the grid is the 'Help Defeat Cancer Project', which uses advanced image processing and pattern recognition techniques to determine the protein expression level in TMAs. The grid speeds up this process dramatically and can analyse 130 years of computation in one day. The long-term goal of the project is to build up a comprehensive library of biomarkers and their expression patterns that can be consulted to find the most appropriate treatment for a specific type of cancer.

The major disadvantage of grid computing is that it is only suitable for CPU-intensive applications, and not for applications which require a lot of data transfer. This is because the data is sent out over standard Ethernet or an Internet connection using TCP/IP as the transfer protocol. TCP/IP divides data up into packets and ensures each packet arrives at its destination before sending out another one. This means the system has a high fault tolerance but at the expense of communication. If there is a vast amount of data to be sent out the network can become a limiting factor and slow down the application. For communication-intensive applications, clusters of FPGAs are a better choice.

1.7.3 Field-programmable Gate Array (FPGA)

An FPGA is a hardware device that slots into the PCI slot of a desktop PC and acts as an additional processing device. FPGAs were invented in 1984 and consist of programmable logic blocks and interconnections. They are typically parallel devices and can contain hundreds of specially-developed processors. The speed increase is not due to their parallel nature but is because the algorithms are programmed in hardware rather than software. Typical applications of FPGAs include medical imaging and computer vision.

ClearSpeed technology is a hardware acceleration technology specifically designed for high-performance computing. It makes a range of accelerator boards that slot into the PCI slot. These boards are not strictly speaking FPGAs but are known as attached processors or co-processors. One product is the ClearSpeed CSX600 Advance Accelerator Board, which provides up to 50 GFLOPS of sustained performance (the equivalent of five workstations) while using just 25 W of power. The accelerator board consists of two CSX600 processors. Each processor has 96 cores, and it is the world's fastest and most power-efficient 64-bit floating point processor. The accelerator board works by accelerating standard math libraries used by common mathematical and image-processing platforms such as MATLAB. This means that software engineers get a performance gain without having to change their code. When a call is made by an application to a ClearSpeed-supported library the computer calculates whether the function is worth offloading to the accelerator board. If it is, the answer is calculated on the ClearSpeed board and sent back to the host computer.

The accelerator board can be used to accelerate a standard desktop PC or can be used in high-performance computers to give a performance boost. The ninth-fastest supercomputer in the world was given a 24% performance boost by adding 360 ClearSpeed Advance boards, without the need to add any additional facilities and only increasing the power requirements by a single percent. ClearSpeed has the potential to be used in the acceleration of analysis from virtual slides in a number of ways. It can be combined with a cluster to give an extra speed boost, or it can be added to a standard desktop PC to create a low-cost imaging analysis solution. In this guise it may potentially integrate into the scanner-controlling software. The benefit of this is that the virtual slides are local to the computational device, thus avoiding the necessity to transfer images to a remote analysis engine, which may be the case for the HPC and grid solutions.

1.8 Conclusions

Current microscopy technology using the conventional microscope has reached a tipping point in its existence. The digital information age and the introduction of virtual microscopy into these domains are set to revolutionize current practices. In today's environment, virtual microscopy has shown itself to have a significant role in pathology applications. Initially we will appreciate ever-increasing uses in the delivery of education, in clinical meetings (multi-disciplinary meetings) and in roles within quality assurance programmes. Eventually virtual microscopy will become a standard tool in routine diagnostics. For now there is a feeling within the pathological/scientific community that virtual microscopy should be regarded as a useful adjunct to conventional microscopy. This will change with time through advances in the scanning equipment, efficient remote delivery of virtual slides and increases in resolution on the viewing platform.

Acknowledgements

We would like to thank all members of staff/students within the CCRCB at Queen's University Belfast who were associated with the projects described. Thanks also to the staff of the Bio-imaging Core Technology Unit at Queen's University for assistance with all issues regarding slide scanning.

Jim Diamond is also a Director of i-Path Diagnostics Ltd., UK.

References

[1] Diamond J, Anderson NH, Thompson D, Bartels PH, Hamilton PW. The use of morphological characteristics and texture analysis in the identification of tissue composition in prostatic neoplasia. Hum Pathol. 2004;35(9):1121–31.

[2] Wang Y, Crookes D, Eldin OS, Wang S, Hamilton PW, Diamond J. Assisted-diagnosis of cervical intraepithelial neoplasia (CIN). IEEE Journal of Selected Topics in Signal Processing, special issue on digital image processing techniques for oncology. 2008 Apr.

[3] Diamond J, Anderson NH, Thompson D, Bartels PH, Hamilton PW. A computer based training system for breast fine needle aspiration cytology. J Pathol. 2002;196: 113–212.

[4] Richards WG. Innovation: virtual screening using grid computing: the screensaver project. Nat Rev Drug Disc. 2002;1(7):551.

2

Cytopathology

Mary Hannon-Fletcher

Lecturer, University of Ulster

2.1 Introduction

The study of cytopathology requires a sound understanding of basic cell biology, biochemistry, physiology and anatomy; all of these subjects detail the normal structure, function and metabolism of the human body. An in-depth understanding of 'normality' is essential before pathology and disease may be determined.

The discipline of cytopathology does not exist in isolation but interacts and impacts on other disciplines, such as those mentioned above. Indeed, many of the histochemical techniques employed today are modifications of methods originally developed for use in biochemistry.

Every book, chapter and paper that deals with cytopathology will acknowledge the pioneering work conducted by Dr George N. Papanicolaou. This chapter is no different!

Dr Papanicolaou published his first work on the menstrual cycle of women, *The sexual cycle in the human female as revealed by vaginal smears*, in 1933 [1]. While researching for this publication he noted abnormal 'cancer' cells in the cervix. It was not until 1939 that the true diagnostic significance of this finding was realized. As a result of his efforts the screening of vaginal smears for cancer detection was introduced in a New York hospital that same year.

This was followed by a long collaboration with Dr Herbert Traut, from the department of Obstetrics and Gynaecology at Cornell; together they demonstrated the diagnostic potential of the vaginal smear. Their findings were published in 1943, as *Diagnosis of Uterine Cancer by the Vaginal*

Advanced Techniques in Diagnostic Cellular Pathology Edited by Mary Hannon–Fletcher and Perry Maxwell

Smear [2], and thus the Papanicolaou (Pap) test was born. Dr Papanicolaou published the first comprehensive scientific cytology text, entitled *Atlas of Exfoliative Cytology*, in 1954 [3], in which he describes in detail the cytological morphology of cells, in health and disease, of all the major organ systems of the human body.

Papanicolaou contributed greatly to our knowledge and understanding of the cytological changes associated with cancer and developed cytology (or cytopathology) from a purely theoretical field into the accepted laboratory discipline we know today.

Cytopathology is ever-expanding, with new techniques developing rapidly hand-in-hand with emerging molecular biological technology. Indeed several of these new techniques have become part of the diagnostic laboratory's routine tests (for example polymerase chain reactions (PCR) in diagnostic cytopathology).

2.2 Basic principles

To understand the terminology employed here and elsewhere it is necessary to start with some definitions. The first is 'cytology', which may be described as the scientific study of the structure and function of cells, while 'cytopathology' (also known as 'cellular pathology') involves the examination of cells in health (usually screening), in disease (diagnostic) and for research. It is very important to distinguish between these three as this will tell us some important information about the specimen. To explain, screening is the examination of specimens from asymptomatic individuals in order to detect pre-malignant or early malignant changes, for example the Cervical Screening Programme. Diagnostic cytopathology is when a patient presents with symptoms and specimens are taken to determine a diagnosis. Primarily the specimens will be assessed to determine if they are malignant or benign, but secondary diagnosis may be made, such as the presence of inflammatory or viral changes. Finally, with the introduction of liquid-based cytology (LBC), which has aided optimal specimen collection and facilitated research to be undertaken more readily with ample well-preserved specimens, the screening and/or diagnostic process is complete.

2.2.1 Specimen collection

Specimen collection is just as important as optimal processing and diagnosis. Cytomorphological assessment by highly-trained individuals via the light microscope is fundamental to diagnostic cytopathology. This principle is important in cytopathology, and it differs here from other branches of laboratory medicine because of the nature of specimen collection. The fundamental principle of cytopathological diagnosis is that the content of a cytology specimen should be representative of the cell population of the target tissue or lesion. This is essential as nonrepresentative specimens result in misdiagnosis (false positive or false negative) or no diagnosis (repeat unrepresentative specimen). All of this plays a role in the sensitivity and specificity of the technique.

2.2.2 Specimen sample collection

There are three main methods employed for specimen collection in cytopathology. They are: exfoliation, abrasion and aspiration. Exfoliated cells shed naturally from the epithelial surface and are present in sputum and urine. Such cells appear as small clusters without order, are spherical and are susceptible to degenerative changes. Abrasion uses physical force to obtain cells, for example brushings from the cervix and bronchus, or specimens obtained by scraping for example the cervix, skin and nipple. Specimens may also be obtained using washings or lavage (isotonic saline is used to wash the site and the fluid collected) from the bronchial tract. Such cells are generally well preserved, in large groups or cohesive aggregates. Finally, specimens may also be obtained by aspiration, for example fine-needle aspiration (FNA) or fine-needle punction (FNP). Most sites are accessible using a fine needle, which collects the cellular material. In order to obtain optimal sampling, radiological imaging is used to locate small, deep, mobile lesions (Figure 2.1). Cells are removed via a fine-bore needle, 19–25-gauge, attached to a syringe with or without suction from solid tissue or even fluid-filled cavities. The needle enters into the lesion, then the syringe plunger is partially withdrawn to create a vacuum and the suspect cells are aspirated. This step may be repeated several times. When sufficient cellular material has been withdrawn the syringe plunger is released, which allows the pressure to equalize, and then the needle is withdrawn and the cellular material is fixed. For more information on FNA I would recommend [4].

2.2.3 Specimen preparation

This differs quite substantially from histology, where tissue specimens have to be fixed, dissected and processed (usually overnight), and sections must

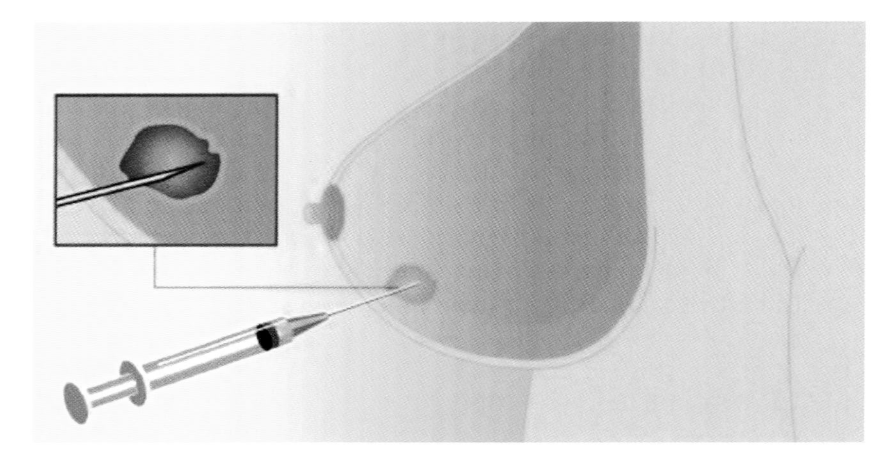

Figure 2.1 Diagrammatic representation of a fine-needle aspirate from the breast. A 19–25 gauge needle is used to aspirate suspect cells from a breast lesion, then the syringe plunger is partially withdrawn to create a vacuum and the suspect cells are aspirated. This step may be repeated several times. When sufficient cellular material has been withdrawn the syringe plunger is released, which allows the pressure to equalize, and then the needle is withdrawn and the cellular material is fixed

be cut on the microtome before slides are prepared for staining. A specimen will arrive in the cytopathology laboratory either already prepared (known as the direct smear method) or as a collection of cells in a fluid suspension, which have to be further processed (the indirect method). Ideally a smear should be evenly spread, uniformly thin and flat, allowing for rapid air-drying and even penetration of fixative in order to permit optimal staining.

2.2.3.1 The direct smear

This involves spreading fresh material across a slide using another slide, pick or spatula, and is usually performed in the clinic or GP surgery by the nurse or clinician (Figure 2.2). The specimen is immediately fixed in alcohol (see Section 2.2.5, Wet fixation) and posted into the laboratory together with the request form.

2.2.3.2 The indirect smear – cell concentration

In this method the cellular material is suspended in fluid, for example saline or transport medium. A cell-concentration procedure (see Section 2.2.4) needs to be performed to increase the yield and to prepare the slides. Cell deposits from washings and urine may not adhere well to glass slides, so an

Figure 2.2 Slide preparation via the direct smear method. The cellular material has been collected on to the spatula. The spatula is then used to spread the fresh material across a slide. It is essential that the material is spread as evenly as possible and immediately fixed

adhesive is employed. Glass slides are pretreated with bovine serum albumin (BSA), poly-l-lysine or amino-propyl-triethoxy-silane (APES). All methods will enhance adherence and maximize the cellular material for examination.

2.2.3.3 Other smear techniques

Touch-imprint preparations may also be used, although they are not very common. This is when a patient presents with an open lesion and the slide is pressed directly on to the suspect area and immediately fixed. Squash preparations have been employed in pre-neurosurgical diagnosis, where this method is used in preference to conventional histology on frozen sections when a rapid diagnosis is required. Finally, processing a clot from a very blood-stained specimen may be required, as during the formation of the clot much of the cellular material may have been sequestered (that is if the clot cannot be dispersed by mechanical means). The clot is then processed as a cell block and is submitted for conventional histology.

2.2.4 Cell concentration techniques

There are four main techniques employed in cytopathology to concentrate cell samples. They are: centrifugation, cytocentrifugation, membrane filtration and cell-block preparation.

2.2.4.1 Centrifugation

This method is suitable for large-volume specimens such as urine, serous effusions and salinated lavage specimens. The procedure is simple: the specimen is transferred to a labelled sterile centrifuge tube and centrifuged to concentrate the specimen. Following centrifugation the supernatant is removed and the precipitate (where the cells have been concentrated) is used to prepare slides, as described in Figure 2.2; however, a glass slide is used to spread the cellular material rather than a spatula.

2.2.4.2 Cytocentrifugation

This method is suitable for small-volume specimens containing little cellular material. The fluid obtained is spun directly on to microscope slides, forming a localized monolayer of cells. As some material is absorbed into the filter paper, very cellular or viscid specimens should not be processed by this technique. This method is often used in preparing samples for subsequent testing such as immunocytochemistry (see Chapter 4).

2.2.4.3 Membrane filtration

This method employs positive pressure or vacuum filtration. There are various types of filter paper, for example cellulose acetate and polycarbonate, each available in a series of pore sizes. This is ideal for large-volume specimens and for large-volume hypocellular specimens as a higher cell yield may be obtained by this method than through centrifugation.

2.2.4.4 Cell-block preparation

This method uses cells that have aggregated into a 'tissue-like' specimen that can be processed as a cell block and cut using conventional histology techniques. Commercial kits are available to prepare such blocks. Alternatively, 1%-cell-culture-grade agarose can be used as a processing medium. This method is suitable for most cell suspensions and has the added bonus that the resulting specimen is suitable for established special staining techniques including immunocytochemistry.

2.2.5 Specimen fixation

Two methods are employed in the cytopathology laboratory: wet and dry fixation.

2.2.5.1 Wet fixation

The most common fixative used in a cytopathology laboratory is alcohol-based, i.e. ethanol. This dehydrates the cytoplasm and coagulates the protein. When specimen preparation has taken place in the laboratory, the slides are immersed in the fixative for a minimum of ten minutes. When the slides have been prepared in the clinic, polyethylene glycol (PEG, or a similar agent) is added to the alcohol. This provides a protective waxy coating when the specimens have to be posted. This coating must be thoroughly removed prior to subsequent fixation (see Chapter 4).

These methods induce a degree of shrinkage in the final preparation, which is why optimal specimen slide preparation is essential.

2.2.5.1.1 Other fixatives Gluteraldehyde and formalin may be used under certain circumstances, along with Carnoy's fluid, which is often used as it lyses erythrocytes and may be helpful in heavily blood-stained specimens. For a more detailed account of fixatives that can be employed in cytopathology I would refer the reader to [5].

2.2.5.2 Dry fixation

This relies on evaporation and has to be rapid for optimal results. It is therefore best to employ forced air movement rather than passive air. This method has a tendency to flatten cells, and they appear larger than when they have been wet-fixed. Post-fixation in methanol is essential to prevent cross-infection from unscreened specimens.

Irrespective of choice, adequate fixation bearing in mind subsequent staining and analytical techniques is essential.

2.2.6 Staining methods

It is worth noting that the prefix cyto- may be replaced with the prefix histo- when the sub-cellular detail rather than the multi-cellular structure is discussed. In many cases some terms may be interchangeable. It follows that most of the techniques employed on tissue (histo-) may also be employed on cytopreparations. For detail-staining protocols, and more information on special staining techniques, I would recommend [5].

2.2.6.1 The Papnicolaou (Pap) stain

The most common stain used in cytopathology is the Pap stain. The differential staining pattern permits prolonged periods of microscopy with minimal eye-strain and it is the stain of choice for gynaecology cervical cytology specimens. It is also used for staining non-gynaecological specimens. However, this is dependent on personal preference and on the availability and number of slides prepared.

2.2.6.2 Romanowsky stains

Romanowsky stains are performed on air-dried preparations and may be automated or rapid manual techniques. They are most commonly used for non-gynaecological specimens.

2.2.6.3 Haematoxylin and eosin (H&E)

The H&E stain can be used on both wet- and dry-fixed specimens. It is often favoured for histology, by those who are familiar with it.

The numerous staining techniques employed in histology may also be employed for cytopathology specimens as required. More examples of special staining techniques are detailed in subsequent chapters.

2.3 Cytodiagnosis

The clinical role of the cytopathologist is to diagnose disease processes in patients. He/she achieves this by interpreting morphological changes in the context of clinical medicine and must therefore remain a clinician in order to be able to relate findings to patient management. The cytopathologist remains in contact with different facets of medicine and is constantly consulted by clinical colleagues through diagnostic work and autopsy findings.

Cytodiagnosis is a very complex process, with the final conclusion being dependent on many factors, including: detailed site-specific knowledge; experience of normality; familiarity with the numerous appearances of disease; awareness of mimics and artefacts; clear understanding of the limitations of the technique; and detailed clinical information.

2.3.1 Morphology

Specimen cellularity provides us with important information about the target organ/tissue. The number and type of cells present give important information about the target tissue. For example, a very cellular specimen may be caused by increased cell activity, known as hyperplasia, which may be either physiological (e.g. breast enlargement) or pathological (e.g. prostate enlargement).

On the other hand, an acellular specimen could mean decreased cell activity (hypoplasia), which may be physiological (e.g. due to atrophy, apoptosis, senescence or degeneration) or pathological (e.g. hypoplastic left heart syndrome). It is the job of the cytopathologist to discriminate between changes that are:

- non-neoplastic
- pre-neoplastic
- neoplastic.

This leads therefore to a very complex process, as the cytonuclear differences between low-grade neoplasia and regenerative or hyperplastic cells may be subtle and/or indistinguishable. Indeed, post-radiotherapy, chemotherapy, viral changes and so on may mimic neoplasia (Figure 2.3).

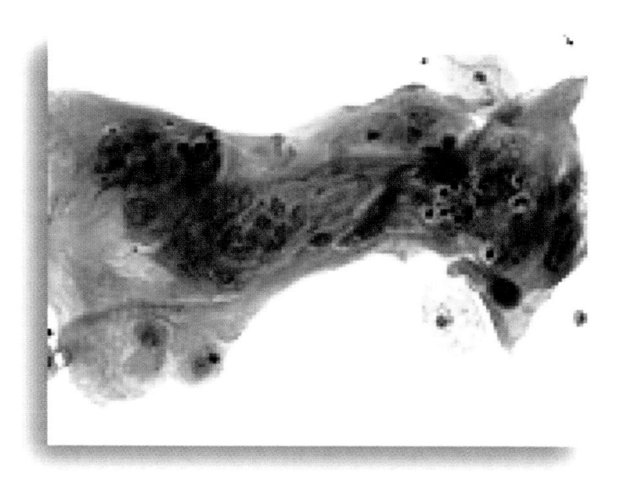

Figure 2.3 Radiotherapy changes. A cervical smear showing typical post-radiation changes, multinucleation, vacuolization, altered chromatin and inflammatory cell engulfment. Magnification X40

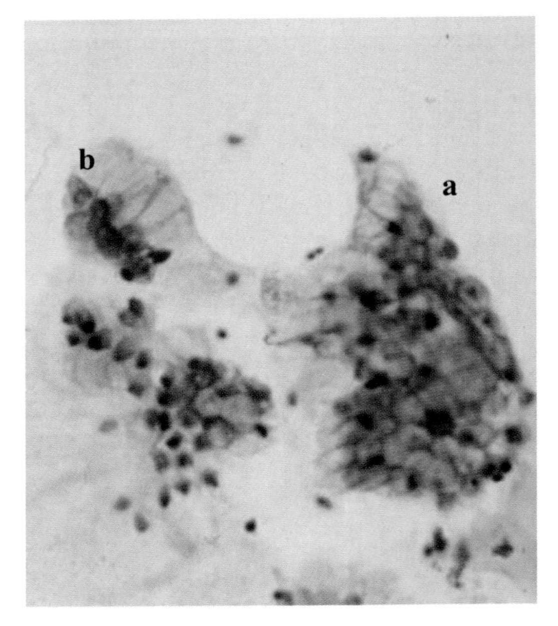

Figure 2.4 Normal glandular epithelium. Typical cytoarchitectural pattern of normal glandular epithelium, regularly arranged in monolayers: a) en face, honeycomb; b) in profile, picket-fence pattern. A cervical smear stained with Pap stain

2.3.1.1 Cytoarchitectural pattern

Architectural features are seldom seen in cytopreparations. However, normal epithelium from many sites retains polarity and intercellular adhesion. Glandular epithelium yields regularly-arranged monolayers: en face, honeycomb; or in profile, picket-fence pattern (Figure 2.4).

2.3.1.2 Nuclear and cytoplasmic morphology

Normal cells are characterized by their morphological uniformity and specific detail in the following:

- nucleus
- cytoplasm
- nuclear:cytoplasmic ratio
- cell size and shape.

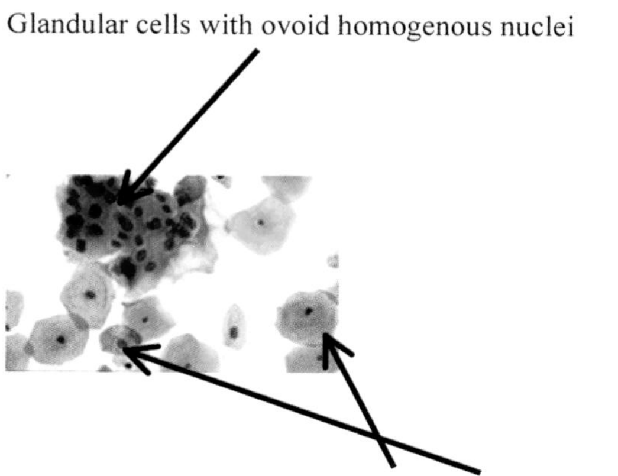

Figure 2.5 Normal squamous and glandular epithelium. A cervical smear stained with Pap stain. Note the nuclei are relatively small compared to the overall volume of the cell in both the squamous and glandular epithelium, although this is clearer in the squamous cells. The nuclei are round or ovoid in shape, and have a smooth nuclear contour and evenly-distributed chromatin, which is homogeneous and finely granular. There is little variation between cells of similar type and maturity

Changes in the morphology of the above are important diagnostic indicators as the nuclear morphology reflects the state of proliferation and reproductive capacity of the cell, and the cytoplasm generally provides an indication of origin, functional state and degree of differentiation.

2.3.1.3 The classical features of a normal mature nucleus

A normal mature nucleus is relatively small compared to the overall volume of the cell; it is round or ovoid in shape, with a smooth nuclear contour and evenly-distributed chromatin, which is homogeneous and finely granular. There is little variation between cells of similar type (Figure 2.5).

2.3.1.4 Nuclear:cytoplasmic ratio

In a two-dimensional cytopreparation, the nuclear size is proportional to the relative nuclear area (cytonuclear index). Poorly-differentiated cells usually possess enlarged nuclei for the same absolute cytoplasmic volume, that is an elevated nuclear:cytoplasmic ratio. This is often difficult for the untrained eye to detect, so I have prepared a series of diagrammatic representations of increasing nuclear:cytoplasmic changes, from normal, through dyskaryosis, to examples associated with cancer/invasion (Figure 2.6).

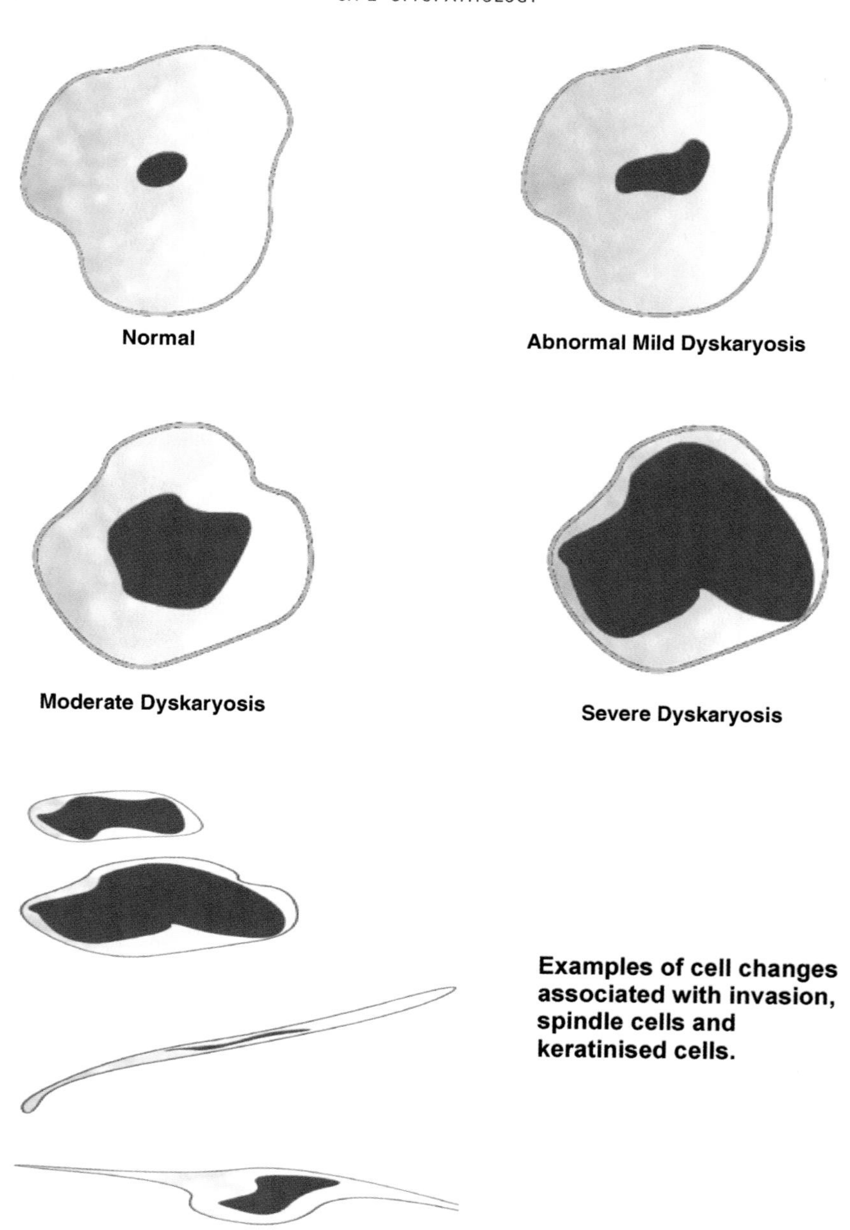

Figure 2.6 Diagrammatic representation of nuclear cytolasmic ratio in a mature cell. Examples of the nuclear:cytoplasmic ratio seen in a normal mature cell through mild, moderate and severe dyskaryosis to cancer/invasion. As the severity of the lesion increases so the nucleus occupies a greater proportion of the cellular volume, until cancer and invasion, where the cytoplasm has almost been lost to a scanty strip or is almost fully keratinised

Figure 2.7 Changes associated with HPV infection. A cervical smear stained with Pap stain showing the classical features of an HPV infection. Typical nuclear clearing (koliocyte), irregular nuclear border and the slightly heterogeneous nuclear chromatin are all diagnostic of HPV infection

2.3.1.5 Background milieu and artefacts

In a traditional smear there will be background details that may or may not be helpful in reaching a diagnosis. The cytopathologist must first determine if these features are cellular or noncellular. For example, there are numerous features that can be found in directly-prepared slides which could be artefact or diagnostic. The smear could be intensely blood-stained with the blood obscuring the cells, making a diagnosis impossible. The same can happen in cases of severe infection, where the agent may be bacteria such as *lactobacillus spp*, fungus such as *Candida albicans*, viral changes such as in koliocytosis (Figure 2.7) (found in human papilloma virus (HPV) infection) or a ground-glass nucleus (found in herpes simplex virus (HSV) infection) (Figure 2.8).

Other agents are often observed, including crystals, mucus, fibrinopurulent, connective tissue stromal content exudates, colloid, basement membrane material, protein-rich fluid; the list is endless, which is another reason why the cytopathologist is required to have such an extensive site-specific knowledge of normality.

2.4 Gynaecological cytopathology

The British Society for Clinical Cytology working party on terminology recommended that the term dyskaryosis was to be used when describing

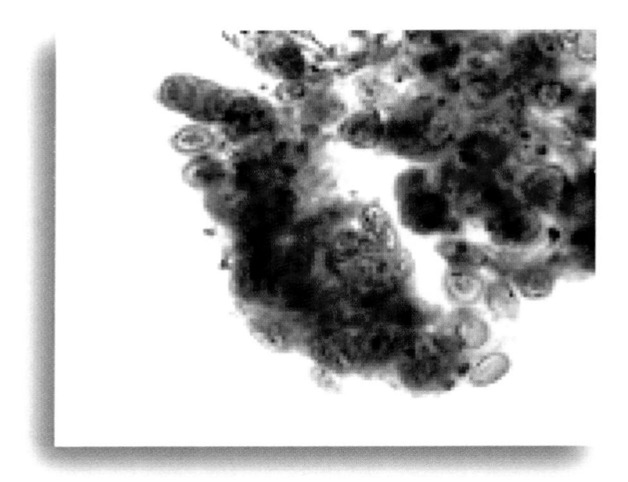

Figure 2.8 Changes associated with HSV infection. A cervical smear stained with Pap stain showing the classical features of an HSV infection. Typical cellular changes such as multinucleation, nuclear moulding, ground-glass chromatin, chromatin margination and intranuclear viral inclusions are all diagnostic of HSV infection

abnormal nuclear morphology. Dyskaryosis comes from the Greek for 'abnormal nucleus', first described by Papanicolaou. The changes are categorized as mild, moderate and severe, which are all pre-invasive and may be correlated with the histological terminology 'cervical intraepithelial neoplasia' (CIN), where mild dyskaryosis broadly relates to CIN 1, moderate to CIN 2 and severe to CIN 3 (Figure 2.9). The classical changes associated with these catagories are grouped into nuclear and cytoplasmic changes, outlined in Table 2.1(a) and (b), respectively. For a detailed description of the morphological changes found in pre- and malignant cells I recommend [6], which is the most detailed volume to be found.

Figure 2.9 (*Continued*) moderate dyskaryosis (dysplasia). CIN 3 changes occupy >2/3 of the epithelium and may involve full thickness (also referred to as carcinoma *in situ*). This stage is equivalent to severe dyskaryosis (dysplasia). CIN 3 is a precursor to invasive carcinoma of the cervix. An important point must be made here: a cytology report of mild or moderate dyskaryosis does not always correlate well with the presence of CIN 1 and CIN 2. A deciding factor is often the quantity of CIN rather than the quality. Histological samples may display CIN 3 changes after a mildly or moderately dyskaryotic cytology smear result. However, the degree of disparity is not usually so great

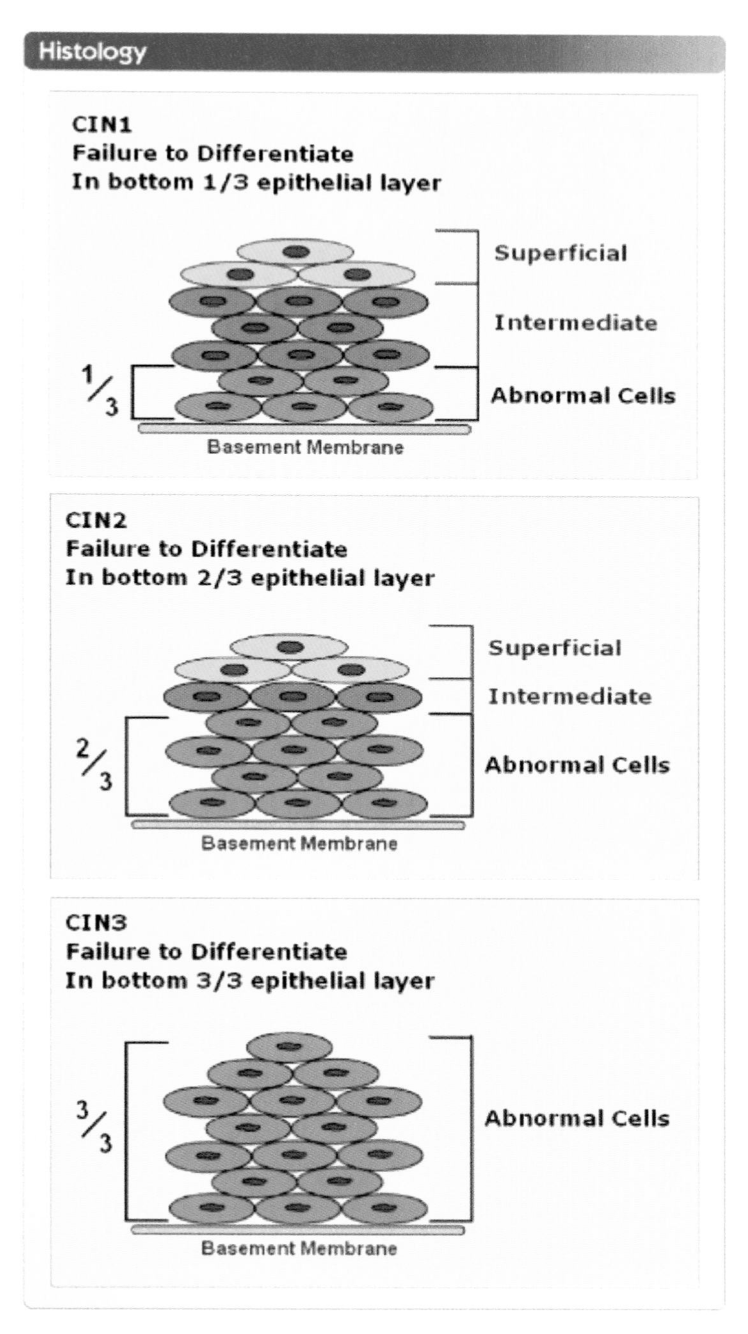

Figure 2.9 Histological classification of CIN 1–3. CIN 1 changes are confined to the basal 1/3 of the epithelium and are equivalent to mild dyskaryosis (dysplasia) in cytopathology. CIN 2 changes are confined within the basal 2/3 of the epithelium and are equivalent to

Table 2.1 Changes associated with a) the cytoplasm and b) the nucleus. The changes listed may be observed alone, usually in cases of mild dyskaryosis/dysplasia, or several may be observed together. In general the more observed together the greater the degree of dyskaryosis. It is however not necessary to see all together for the lesion to be severe

a)

Cytoplasmic Features

Variation in size and shape of similar cells – anisocytosis

Abnormal cells may have:

larger cytoplasmic volume
smaller cytoplasmic volume
nuclear:cytoplasmic ratio affected
ultrastructural composition of the cytoplasm exerting a major influence on staining reactions, e.g. mitochondria, golgi, lipid, carbohydrate & hormones, etc.
keratinization – squamous differentation
cytoplasmic moulding
cell engulfment.

Degenerative changes:

swelling
vacuolation
loss of integrity of plasma membrane – cytolysis.

b)

Nuclear Features

Increased nuclear DNA content results in:

dense nuclear staining – hyperchromasia
irregular, coarse and clumped chromatin
thickened nuclear envelope – karyotheca
variation in size and shape of nuclei between cells – anisokaryosis or anisonucleosis
nuclear membrane becoming irregular, with grooving, indentation or crenation
multi-nucleation being observed.
nucleoli
normal nuclei containing small, discrete nucleoli composed of RNA and associated proteins
both multi-nucleolation and macro-nucleoli observed in proliferative states – both neoplastic and non-neoplastic.

Degenerative chromatin may appear as:

a dense, contacted mass
dense fragments
undergoing dissolution.

2.4.1 The Cervical Screening Programme

The Cervical Screening Programme was introduced in England in 1964 [7]. The National Health Service Cervical Screening Programme (NHSCSP), as we know it today in the United Kingdom, was introduced in 1988 [8]. The new programme employed a computerized call-recall system, and the Department of Health, UK recommended that all women between the ages of 20 and 64 years should be invited for screening every 3–5 years.

The programme had several important goals: to reduce the mortality from cervical cancer; to reduce the incidence of invasive cancer; and to identify and treat as many women as possible who develop CIN 3.

These screening programmes have been very successful in reducing both incidence and mortality from the disease by early detection and treatment of the pre-cancerous changes in the cervix [9]. In the United States, death rates from cervical cancer reduced by 74% between 1955 and 1992 [10], while in the United Kingdom approximately 1–2% of smears were reported as showing moderate or severe dyskaryosis [11]. Figures from Cancer Research UK [12] show that mortality rates resulting from cervical cancer increase with age and women over the age of 75 have the highest percentage of cancer deaths, while women under 35 years of age account for only 6% of cervical cancer deaths [12]. Women receiving an abnormal report are followed up with the relevant referral procedure according to the degree of abnormality (Table 2.2).

However, even with the obvious success of established cervical screening programmes such as the NHSCSP, deaths still occur from cervical cancer. In the United Kingdom in 2006 there were 949 deaths from cervical cancer, which equates to a European age-standardized death rate of 2.4 per 100 000 females (Table 2.3). Therefore, research is underway to develop new and improved technologies/tests to further reduce disease incidence and mortality. In this molecular age there have been numerous new technologies developed, including the introduction of vaccinations against high-risk strains of HPV, molecular HPV testing, LBC, automated screening technologies and immunocytochemical staining techniques.

2.4.2 Cervical cancer

Cervical cancer is the fifth most deadly cancer in women worldwide [13]. Early stages of this disease can be totally asymptomatic. The most commonly-noted symptom is vaginal bleeding or spotting at abnormal times such as between periods, post coitus or post menopause. Other symptoms include discomfort or moderate pain during intercourse and vaginal discharge. When the disease reaches an advanced stage symptoms can include weight loss, pelvic pain, loss of appetite, fatigue, heavy vaginal bleeding and back pain [14]. There are two main types of cervical cancer: squamous cell carcinoma, accounting for approximately 85% of all cases,

Table 2.2 Follow-up procedure following an abnormal Pap smear result. Management of the cytology result is dependent on the degree of severity of dyskaryosis and its persistence. In the United Kingdom there is a recall system in place and if required patients may undergo a procedure called colposcopy. During colposcopy small tissue samples are collected for histological diagnosis of CIN

Smear Result	*Inadequate*	*Borderline Dyskaryosis*	*Mild Dyskaryosis*	*Moderate Dyskaryosis*	*Severe Dyskaryosis*
Description of Result	Insufficient material present or poorly spread/fixed	Nuclear changes that are not normal are present. Unsure whether the changes represent dyskaryosis (*5–10% of all smears are borderline or mild*)	Nuclear abnormalities that are indicative of low-grade CIN* (*5–10% of all smears are borderline or mild*)	Nuclear abnormalities reflecting probable CIN 2* (*~1% of all smears*)	Nuclear abnormalities reflecting probable CIN 3 (*~0.6% of all smears*)
	Vision of cells obscured by debris (*~ 9% of all smears*)				
Action Required	Repeat the smear immediately	Repeat smear in six months	Repeat smear in six months	Refer to colposcopy	Refer to colposcopy or (rarely) make urgent referral to gynaecological oncologist (if invasive carcinoma is suspected)
	After three consecutive inadequate results, refer for colposcopy	Most smears will have reverted to normal	Most smears will have returned to normal		

		After three consecutive normal smears, return to normal recall	After three consecutive normal smears, return to normal recall		
		If abnormality persists (three times) or worsens, refer for colposcopy	Refer for colposcopy if changes persist on two occasions		
		If in a ten-year period there are three borderline or more severe results, refer to colposcopy	If in a ten-year period there are three moderate results, refer to colposcopy		
Colposcopy					
Results			CIN 1 confirmed histologically	CIN stage confirmed histologically	CIN 3 confirmed histologically
Action			Management options not clear (continue to watch and wait, or treat?)	Treat to remove areas of abnormal cells	Treat to remove areas of abnormal cells

*As already mentioned in Figure 2.9, an important point must be made here: a cytology report of mild or moderate dyskaryosis does not always correlate well with the presence of CIN1 and CIN2. A deciding factor is often the quantity of CIN rather than the quality. Histological samples may display CIN3 changes after a mildly or moderately dyskaryotic cytology smear result. However the degree of disparity is not usually so great [12].

Table 2.3 European age-standardized mortality from cervical cancer by age group, England and Wales. During the second half of the twentieth century the death rate from cervical cancer for women aged 55–64 dropped by nearly 80%, from 30.0 per 100 000 in 1950–52 to 6.2 per 100 000 in 1998–2000 [12]

	Age Group				
Rate per 100 000 females	*25–34*	*35–44*	*45–54*	*55–64*	*65 +*
1950–52	1.8	7.4	18.0	30.0	33.7
1998–2000	1.3	4.0	5.2	6.2	11.8
Percentage Decrease	27.0	47.0	71.0	79.0	65.0

and glandular cell carcinoma (adenocarcinoma), accounting for 10–15% of cases.

Records indicate that there are approximately 470 000 new cases of cervical cancer and 233 000 deaths annually, with approximately 80% of all deaths from cervical cancer occurring in poor countries [15]. Figures from the United States show that most of the cervical cancers found annually (approx. 11 000) arise in women who have either never had a Pap smear, or have not had one in the previous five years [16].

2.4.2.1 Human papilloma virus (HPV)

HPVs are of the *Papovaviridae* family, which consists of small DNA viruses 55 nm in diameter with a non-enveloped icosahedral outer coat surrounding a circular genome of double-stranded DNA of approx. 8000 bp (Figure 2.10). Their classification is based on DNA sequence differences in the coding region of certain proteins: early genes E1, E2, E4, E5, E6 and E7, and late genes L1 and L2. To date, over 130 HPV types have been described. The early genes are responsible for DNA replication, transcriptional regulation and transformation. The late genes control the formation of the capsid coat. The early gene products E6 and E7 encode the major transforming proteins that are capable of inducing cell proliferation and immortalization by binding to the tumour-suppressor gene products p53 and the phosphorylated retinoblastoma (pRB) protein [17].

Under normal cellular conditions, pRB and p53 regulate cell growth. When infected by HPV, these functions may be disrupted and so cellular transformation may occur. In order to progress to cervical cancer, HPV infection with type 16 or 18 is essential; however, other cofactors have been recorded. These include the use of oral contraceptives for five or more

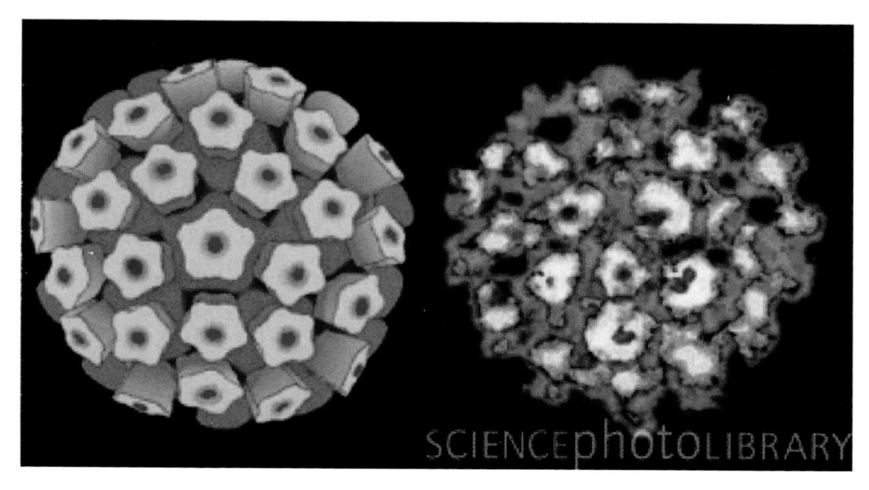

Figure 2.10 Computer artwork (left) and coloured transmission electron micrograph (TEM, right) of HPV. Their protein coats, or capsids (purple, seen best in the artwork), enclose their genetic material. The capsids are studded with surface proteins (blue)

years, smoking, high parity (five or more full-term pregnancies) and previous exposure to other sexually-transmitted diseases such as chlamydia trachomatis and HSV type 2 [18].

Infection is initiated when the virus gains access to the basal epithelial cells. This usually occurs via a lesion of minor trauma (during sexual intercourse) allowing the virus access to the target cells near or at the cervical transformation zone. The viral life cycle is linked to keratinocyte differentiation, and viral replication leads to koilocytosis, nuclear enlargement, multinucleation, dyskeratosis and even CIN. Integration into the host chromosome correlates with lesion grade. It is rare in CIN 1 and very common in CIN 3 and cervical carcinomas [14–17].

2.4.2.2 HPV and cervical cancer

Since the advent of recombinant DNA technology and the cloning of the HPV types, the association between HPV, CIN and cancer has been confirmed [15–20]. Compelling evidence, both epidemiological and molecular, indicates that persistent HPV infection is a pivotal step in the development of cervical cancer. However, only high-risk oncogenes can induce cancer and HPV types 16, 18, 45 and 31 most frequently [17–22] are linked to almost 70% of cervical cancers worldwide [23]. The association between HPV and cervical squamous cell carcinoma is higher than the association between smoking and lung cancer [20].

2.4.2.3 Prevalence of genital HPV

Most women infected with genital HPV will not suffer complications from the virus as host immunity can clear most low-risk oncogene HPV infections, and as such obtaining exact figures is very difficult. However, the Center for Disease Control in the United States has estimated that by the age of 50 more than 80% of American women will have contracted at least one strain of genital HPV [24].

2.4.3 Limitations of cervical screening

There are a number of limitations to the current Cervical Screening Programme. It is generally accepted that the false-negative rate in manual screening may be as high as 10% [25, 26]. The sensitivity of a single cervical smear is assessed at 50% [27] and reports from Northern Ireland show that 10% of women have to return for a repeat smear, 10% receive an inadequate result, 2% have three inadequate results and require referral for colposcopic assessment, and 4.5% receive a borderline result [28]. Obviously no test is perfect and there are deficiencies and inherent limitations in this method. Evaluation of the cervical cells is performed manually; this is time-consuming, laborious and prone to human error. A false-negative result, which is when a negative report is issued despite an abnormal lesion being present on the cervix, can be caused by a number of factors relating to preparation and sampling issues, as detailed in Table 2.4.

Table 2.4 Sampling errors using traditional Pap smear collection. Some examples of the types of sampling error that have resulted in the low sensitivity of the traditional Pap smear

Smear Test Deficiencies	
Collection of the cell specimen	Wooden spatula: not flexible, thus uneven spread and distribution
	Smears too scanty: may be due to atrophy/cytolysis
	Not all cells collected are transferred to the slide
Site	The cells must be collected from all of the cervical regions (exo/endo-cervix and squamo-columnar junction)
Blood and inflammation	Cells obscured by blood and exudate (even other epithelia's)
Smear preparation	Poor fixation
	Air-dried artefacts
	Suboptimal staining
Human error	False negative
	False positive

initiating a host immune response against the two oncogenes, established tumours may be eradicated [40]. Finally, the development of HPV vaccines providing protection against a broader range of HPV types is currently underway [41].

2.4.5.2 Mechanism of action

The preventative HPV vaccines, Gardasil and Cervarix, are based on hollow virus-like particles (VLPs) assembled from recombinant HPV coat proteins that target the two most common high-risk HPVs, types 16 and 18. In addition, Gardasil targets HPV types 6 and 11, which together currently cause about 90% of all cases of genital warts [42].

Initial infection is prevented as Gardasil (HPV 6, 11, 16 and 18) and Cervarix (HPV 16 and 18) are designed to elicit virus-neutralizing antibody responses against the HPV types they contain. The protective effects of the vaccines are expected to last a minimum of 4.5 years after the initial vaccination [43]. Since the vaccines only covers some high-risk types of HPV, experts still recommend regular Pap smear screening even after vaccination.

2.4.5.3 Safety

The vaccines are reported to have only minor side effects, such as soreness around the injection area, and are considered to be safe. They do not contain mercury, thimerosal, or live or dead virus, only VLPs, which are incapable of reproducing in the human body [44]. Merck, the manufacturer of Gardasil, will continue to test women who have received the vaccine to determine the vaccine's efficacy over the period of a lifetime.

2.4.5.4 UK pilot study

A pilot study conducted in Manchester, ($n = 2800$) among school girls aged 12–13 years reported that the overall uptake of the vaccines was 71% for the first dose and 69% for the second [45]. The authors report a lower uptake among girls from minority groups and less affluent backgrounds. A higher uptake is required in order that the vaccines are totally effective. In addition, to be cost-effective an uptake of more than 80% is required [46]. However, the results of the pilot were welcomed by Cancer Research UK [47].

2.4.5.5 Vaccine implementation

In September 2008, the Cervarix HPV vaccine was introduced into the United Kingdom's immunization programme. The programme began by immunizing 12–13-year-old girls in school. This will be followed by a 'catch-up' group of 17–18-year-old girls starting in autumn 2009. This catch-up campaign will offer to vaccinate girls aged between 16 and 18 years from 2009, and girls aged between 15 and 17 years from 2010. By the end of the catch-up campaign, all girls under 18 will have been offered the HPV vaccine. This vaccine requires a three-dose schedule at 0, 1 and 6 months [45].

It is hoped that the vaccination programme will prevent 70–80% of cases of cervical cancer [48], however, as with all vaccination programmes, its success depends on uptake.

It will be many years before the vaccination programme has an effect upon cervical cancer incidence, so women are advised to continue to attend regular Pap-test call backs.

2.5 Conclusions

Very little has changed since Papanicolaou first collected vaginal samples, stained them and noted the presence of abnormal cells which he concluded originated in the cervix. The last ten years, however, have seen major changes in how cervical smears are sampled and preserved, and slides prepared.

The Cervical Screening Programmes have greatly reduced the incidence of cervical cancer deaths. However, where there are no such programmes, deaths continue to rise. Perhaps this was and is the most effective method of preventing cervical cancer deaths. Unfortunately, in order to be completely effective, uptake is required to be at least 80% and this level has not been attained in the developing world.

The recent introduction of LBC has improved the method of sample collection and allowed additional tests to be performed on well-preserved specimens. It has also allowed additional research to take place, which will add to our knowledge in the coming months and years.

The very recent introduction of the HPV vaccine holds great promise for the future prevention of cervical cancer and deaths. Again, in order for this to be effective, uptake of approx. 80% is ideal; only time will tell if this level can be obtained and then maintained.

Finally, the majority of the methods described above are effective in the developed world. Our focus now needs to be directed to the developing countries where cervical cancer deaths are on the increase. Perhaps the research facilitated by LBC and the new vaccine production will be able to reduce the death rate in the developing world, in a way that the screening programmes, to date, cannot.

Acknowledgements

Figures 2.3,2.8 and 2.11 are reproduced with the kind permission of Cytyc Corporation. Copyright © 1997–2004 Cytyc Corporation.

Figure 2.10 courtesy of Dr Linda Stannard, UCT Science Photo Library.

Figures 2.1,2.6 and 2.9 are reproduced by the kind permission of the University of Ulster, Institute of Lifelong Learning.

Useful Web sites

American Cancer Society: http://www.cancer.org

Cancer Research UK: http://www.cancerresearchuk.org

Cervical Screening Programme UK: http://www.cancerscreening.nhs.uk

SurePath LBC: http://www.bd.com/tripath/products/surepath/index.asp

Cytyc Corporation, ThinPrep: http://www.thinprep.com

National Institute of Clinical Excellence (NICE): http://www.nice.org.uk

Merck: http://www.merck.com.

References

[1] Papanicolaou GN.The sexual cycle in the human female as revealed by vaginal smears. American Journal of Anatomy, Wistar Institute of Anatomy and Biology; 1933.

[2] Papanicolaou GN, Traut HF. Diagnosis of uterine cancer by the vaginal smear. New York: The Commonwealth Fund; 1943.

[3] Papanicolaou GN. Atlas of exfoliative cytology. Published for the Commonwealth Fund by Harvard University Press, Oxford University Press; 1954.

[4] Kocjan G. Fine needle aspiration cytology: diagnostic principles and dilemmas. Springer-Verlag Berlin and Heidelberg: GmbH & Co. K; 2005.

[5] Bancroft JD, Gamble M. Theory and practice of histological techniques. 5th ed. W B Saunders Co.; 2002.

[6] Koss LG, Melamed MR. Diagnostic cytology and its histopathologic bases. 5th ed. Lippincott Williams and Wilkins; 2005.

[7] Farmery E, Gray J. Report of the first five years of the NHS cervical screening programme. Oxford: National Co-ordinating Network; 1994.

[8] Department of Health and Social Services. Health services management: cervical cancer screening. Health Circular HC. 1988;88:1.

[9] http://www.cancerscreening.nhs.uk/cervical/index.html [accessed 2008 Sep 27].

[10] http://www.cancer.org [assessed 2008 Sep 27].

[11] Patrick JJ. Achievable standards, benchmarks for reporting, and criteria for evaluating cervical cytopathology. 2nd ed. NHSCSP Publication No. 1; 2004.

[12] http://info.cancerresearchuk.org/cancerstats/types/cervix/mortality/ [accessed 2008 Sep 27].

[13] World Health Organization (February 2006). Fact sheet No. 297: cancer. Available from: http://www.who.int/mediacentre/factsheets/fs297/en/index.html [accessed 2008 Sep 27].

[14] Human papilloma virus and cervical cancer: where are we now? BJOG. 2001; 108: 1204–13.

[15] Cervical Cancer Action, funded by the Rockefeller Foundation. http://www.cervicalcanceraction.org/whynow/intro.php [accessed 2008 Sep 27].

[16] National Cancer Institute SEER fact sheet on cervical cancer. http://seer.cancer.gov/statfacts/html/cervix.html [accessed 2008 Sep 27].

[17] Burd EM. Human papillomavirus and cervical cancer. Clin Microbiol Rev. 2003;16:1–17.

[18] Bosch, FX. Iftner, T.The aetiology of cervical cancer. NHSCSP publication No. 22; 2005.

[19] Walboomers JM, Jacobs MV, Manos MM, Bosch FX, Kummer JA, Shah KV, Snijders PJ, Peto J, Meijer CJ, Muñoz N. Human papillomavirus is a necessary cause of invasive cervical cancer worldwide. J Pathol. 1999;189:12–19.

[20] Franco EL. Cancer causes revisited: human papillomavirus and cervical neoplasia. J Natl Cancer Inst. 1995;87:779–80.

[21] Clifford GM, Smith JS, Plummer M, Muñoz N, Franceschi S. Human papillomavirus types in invasive cervical cancer worldwide: a meta-analysis. Br J Cancer. 2003;88:63–73.

[22] Schiffman M, Castle PE, Jeronimo J, Rodriguez AC, Wacholder S. Human papillomavirus and cervical cancer. Lancet. 2007;370:890–907.

[23] Hanz S, Alain S, Denis F. Human papillomavirus prophylactic vaccines: stakes and perspectives. Gynaecol Obstet Fertil. 2007;25:176–7.

[24] Planned Parenthood. In fact, the lifetime risk for contracting HPV is at least 50 percent for all sexually active women and men, and, it is estimated that, by the age of 50, at least 80 percent of women will have acquired sexually transmitted HPV (CDC, 2004; CDC, 2006). Available from: http://www.plannedparenthood.org/issues-action/std-hiv/hpv-vaccine/reports/HPV-6359.htm [accessed 2008 Sep 27].

[25] Koss LG. The papanicolaou test for cervical cancer detection. A triumph and a tragedy. JAMA. 1989;261:737–43.

[26] Ducatman BS, Wang H. The Pap smear. United Kingdom: Arnold Publishers; 2002.
[27] NICE Technology Appraisals in Progress. LBC cervical screening review. Available from: http://www.nice.org.uk/article.asp?a=65850 [assessed 2008 Sep 30].
[28] McGoogan. Morphology: comparison of liquid based verses conventional smears. The Cytopathology Lecture, IBMS Congress Birmingham; 2001.
[29] SurePath. http://www.bd.com/tripath/products/surepath/index.asp [accessed 2008 Sep 27].
[30] Cytyc Corporation, Marlborough, MA 01752. ThinPrep cervical cancer. Available from: http://www.thinprep.com/cervical-screening/hpv.html [accessed 2008 Sep 27].
[31] Moss SM, Gray A, Marteau T, Legood R, Henstock E, Maissi E.Evaluation of HPV/LBC cervical screening pilot studies. Report to the UK Department of Health. Revised 2004. Available from: http://www.cancerscreening.nhs.uk/cervical/evaluation-hpv-2006feb.pdf [accessed 2008 Sep 27].
[32] Strander B, Andersson-Ellstro A, Milsom I, Raodberg T, Ryd W. Liquid-based cytology versus conventional papanicolaou smear in an organized screening program: a prospective randomized study. Cancer (Cancer cytopathology). 2007;111(5):285–91.
[33] Ronco G, Cuzick J, Pierotti P, Cariaggi MP, Dalla Palma P, Naldoni C, Ghiringhello B, Giorgi-Rossi P, Minucci D, Parisio F, Pojer A, Schiboni ML, Sintoni C, Zorzi M, Segnan N, Confortini M. Accuracy of liquid based versus conventional cytology: overall results of new technologies for cervical cancer screening: randomised controlled trial. BMJ. 2007;335(28):1–7.
[34] Denton K. Liquid based cytology in cervical cancer screening is as sensitive as conventional cytology, and has other advantages. BMJ. 2007;335:1–2.
[35] NHS Information Centre. Cervical screening programme. 2006 Dec. Available from: www.ic.nhs.uk/pubs/csp0506 [accessed 2008 Sep 27].
[36] Williams AR. Liquid-based cytology and conventional smears compared over two 12-month periods. Cytopathology. 2006;17:82.
[37] National Cancer Institute. Human papillomavirus (HPV) vaccines: Q & A. Available from: http://www.cancer.gov/cancertopics/factsheet/risk/HPV-vaccine [accessed 2008 Sep 27].
[38] Merck. http://www.merck.com/newsroom/press_releases/financial/2007_0723.html [accessed 2008 Sep 27].
[39] Reuters. Glaxo prepares to launch Cervarix after EU okay. Available from: http://www.reuters.com/article/governmentFilingsNews/idUSL2446805720070924 [accessed 2008 Sep 27].
[40] Roden RB, Ling M, Wu TC. Vaccination to prevent and treat cervical cancer. Hum Pathol. 2004;35(8):971.
[41] Medical College of Georgia, Science Daily. New HPV vaccine under study. Available from: http://www.sciencedaily.com/releases/2007/11/071119113902.htm [accessed 2008 Sep 27].
[42] Lowy DR, Schiller JT. Prophylactic human papillomavirus vaccine. J Clin Invest. 2006;116(5):1167–73.
[43] Harper D, Franco EC, Wheeler A, Moscicki B, Romanowski C, Roteli-Martins D, Jenkins A, Schuind S, Clemens C, Dubin G. Sustained efficacy up to 4.5 years of a

bivalent L1 virus-like particle vaccine against human papillomavirus types 16 and 18: follow-up from a randomised control trial. Lancet. 2006;367:1247–55.

[44] Information from CDC and FDA on the safety of Gardasil vaccine. Available from: http://www.fda.gov/Cber/safety/gardasil071408.htm [accessed 2008 Sep 30].

[45] Brabin L, Roberts SA, Stretch R, Baxter D, Chambers G, Kitchener H, McCann R.Uptake of first two doses of human papillomavirus vaccine by adolescent schoolgirls in Manchester: prospective cohort study. BMJ. 2008;336:1056–58.

[46] Kim J. Human papillomavirus vaccination in the UK. BMJ. 2008;337:303–4.

[47] Cancer Research UK. 'Encouraging' uptake in HPV vaccination trial. 2008. Available from: http://info.cancerresearchuk.org/news/archive/newsarchive/2008/april/18571762 [accessed 2008 Sep 27].

[48] Kane M. Human papillomaviruses (HPV) vaccines: implementation and communication issues. J Fam Planning Reprod Health Care. 2008:3–5.

3

Flow Cytometry

Ian Dimmick

Flow Cytometry Core Facility Manager, Institute of Human Genetics, Bioscience Centre, International Centre for Life

3.1 Introduction

Flow cytometry can be defined as a semi-automated procedure for the interrogation of single cells in a continuous fluid stream, enabling the derivation of simultaneous measurements of multiple extra- and intra-cellular characteristics. The objective of flow cytometry is very simple: to measure by quantitation of photon release, constituents of the membrane, cytoplasm and nucleus of a particular cell or group of cells.

For simplicity, a model of peripheral blood leucocytes can be used to demonstrate some of the more basic functions of the flow cytometer. In the first instance we will look at what is perhaps the easiest, but still an important, aspect of flow cytometry, which is isolating the cells of interest electronically.

The sample of whole blood must first be depleted of the red cells, which can easily be done by using a hypotonic red-cell lysing solution such as ammonium chloride. The effect of this treatment is to hypotonically burst the red-cell membrane and leave the leucocytes unharmed and ready for analysis. This is an important process as red cells will outnumber leucocytes by a factor of approximately 1000 : 1, making the analysis of the leucocytes much more difficult.

Analysing the sample of leucocytes found within a lysed blood sample with the flow cytometer, without the addition of any antibodies or dyes, shows a typical distribution as seen in Figure 3.1.

Advanced Techniques in Diagnostic Cellular Pathology Edited by Mary Hannon–Fletcher and Perry Maxwell

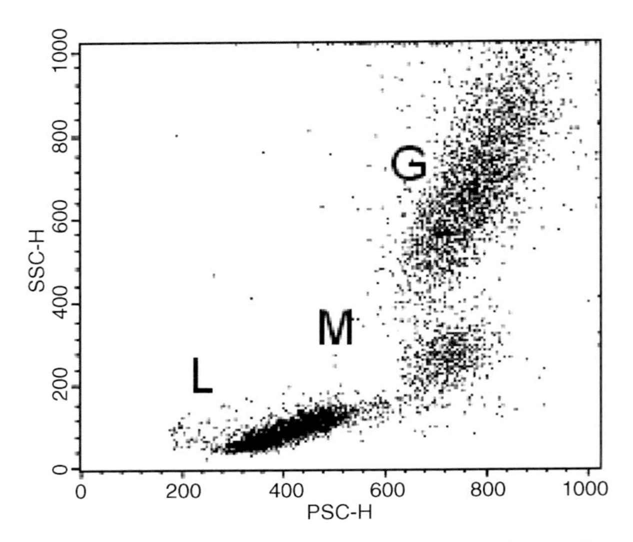

Figure 3.1 Scatter properties of lymphocytes (L), monocytes (M), granulocytes (G) from a lysed sample of whole blood

The three populations of cells described above may not all be of interest to the analysis. An electronic 'gate' therefore can be applied to isolate the lymphocytes from the monocyte and granulocyte populations, which in this case are not of analytical interest. This leads to only the lymphocyte population being observed within the gated histogram, simplifying and increasing the relevance of the statistical analysis (Figure 3.2).

Cell surface and cytoplasmic measurements are usually of cellular antigen expression, where the most commonly used probes are monoclonal antibodies directed to specific antigens on or in the cell; for example, CD3 is a typical surface marker for T cells, while CD22 is a typical cytoplasmic marker for early B cells. Dyes may be used that are specific for DNA, such as Hoechst 33342; RNA, such as Pyronin-y; or other intracellular constituents, such as Indo-1 for calcium.

Multiple antibodies can be used simultaneously within the same sample, each antibody possessing a spectrally different attached fluorescent reporter molecule. These reporter molecules can be made up of various fluorochromes and visualized based on characteristic wavelength emission spectra, as in Figure 3.3.

The flow cytometer is designed to be able to detect the spectral differences of each of these reporter fluorochromes by the use of optical filters, enabling the differentiation of each antibody to specific antigens on the same cell. The resultant data will give an antigenic profile of each cell, predetermined by the specificity of the antibodies used. In the case of the

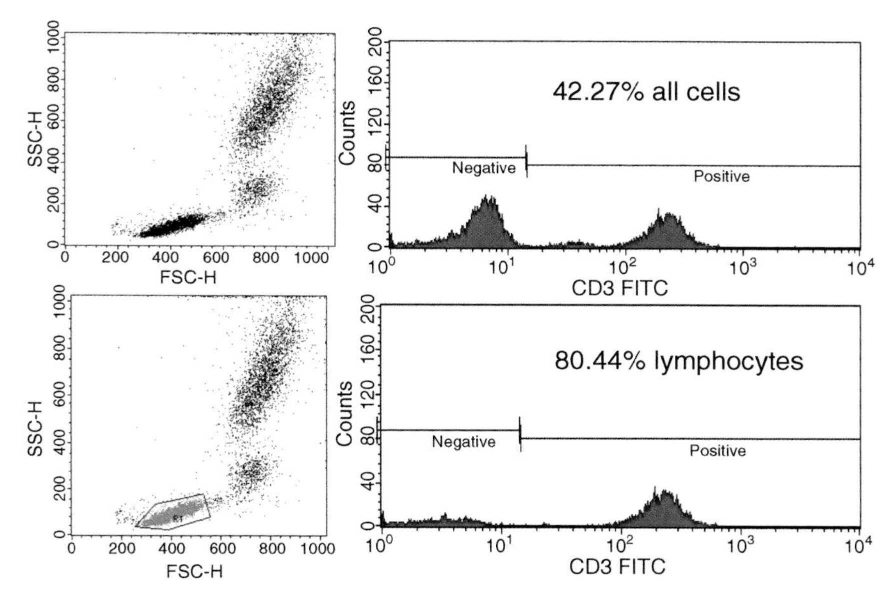

Figure 3.2 Plotting CD3 positivity. CD3 on all cells is 42.27%. Some weak binding to the right of the negatives could be nonspecific monocyte binding of the CD3. Below is an electronic gate placed on to the lymphocytes (R1). The cells from this gate are used in the histogram analysis, showing 80.44% of lymphocytes positive for CD3. The monocyte reaction is eliminated from the analysis

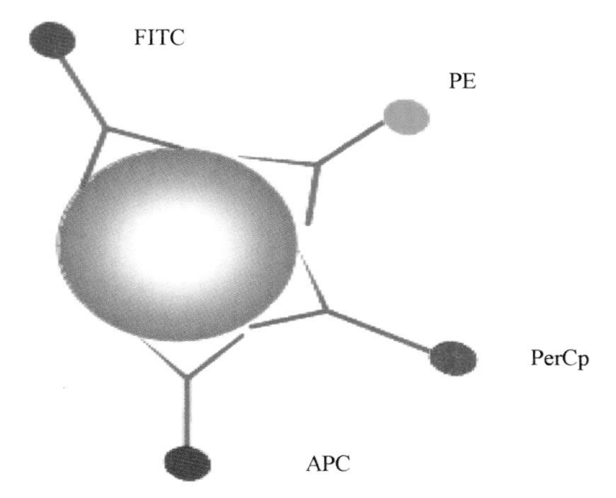

Figure 3.3 Diagrammatic representation of fluorescein isothiocyanate (FITC), phycoerythrin (PE), peridin chlorophyl protein (PerCp) and allophycocyanin (APC) conjugated antibodies attached to the surface of a target cell: FITC excitation 488 nm, emission 525 nm; PE excitation 488 nm, emission 575 nm; PerCP excitation 488 nm, emission 660 nm; APC excitation 635 nm, emission 660 nm

dyes, these are fluorochromes in their own right and will bind to the cell constituents specific to each.

The objective then, once a sample of interest has been obtained from the patient, is to isolate the cells of interest (either electronically by gating, or physically by the use of density-gradient centrifugation), using appropriate probes to identify the target molecules, interrogate the samples by the flow cytometer and acquire the resultant data. One must make sure that the correct excitation sources are used with respect to the probes and that the appropriate detectors are present to measure the emitted fluorescence. Once all this is in place, the data is digitized by the computer and displayed as frequency histograms and dot plots.

3.2 Sample preparation

Good sample preparation is paramount in achieving reliable and accurate results. The objective is to prepare a single cell suspension representative of the sample under investigation, avoiding, or at least minimizing, any losses of cell population due to processing artefacts such as centrifugation or density-gradient separation. Flow cytometry can investigate almost any cell type, but one special example is that of whole-blood analysis.

Leucocytes can be analysed from a sample of whole blood without the necessity for density-gradient centrifugation. The process involves staining the sample of blood either before or after red-cell lysis, or with a vital DNA dye. The use of a vital dye is sometimes very useful in instances when red-cell membranes become resistant to lysis, such as in HIV infection or certain haemoglobinopathies. Red-cell lysis is achieved by adding a hypotonic solution that will swell and then eventually burst the red cell, while leaving the leucocytes relatively unharmed. The alternative is to use vital nuclear dyes such as DRAQ5 (Biostatus), Cytrac Orange (Biostatus) or Vybrant (Invitrogen); the dye binds to the DNA of the nucleated cells after permeating their membranes. This enables the gating of dye-positive cells for examination within the analysis histograms, by excluding any dye-negative non-nucleated cells (i.e. red cells). This can have the added benefit when dealing with antigens of being sensitive to lysing reagents. One example of this strategy can be seen in Figure 3.4, which demonstrates the use of DRAQ5 to visualize the nucleated cells in a sample of peripheral blood without using lysis reagents.

It is very important when adding antibodies to a sample to ensure that the number of cells within the sample tube is known. This enables the addition of excess antigen, thereby avoiding weak staining. This is sometimes difficult as not all manufacturers specify an antibody volume per cell

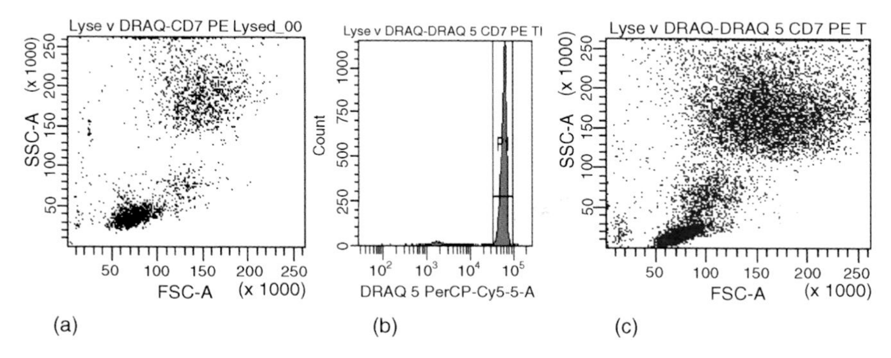

Figure 3.4 Whole-blood sample following: (a) lysis using ammonium chloride; (b) a sample not lysed but DRAQ5 stained; (c) thresholded whole blood expressed as FSC V SSC in the last dot plot. FSC, forward scatter; SSC, side scatter

concentration. In this case the user must titre the antibody for maximal fluorescent intensity on a known positive cell population at a defined cell concentration (usually 1×10^6/ml).

3.3 Principles of the flow cytometer

3.3.1 Introduction of the sample to the flow cell

The flow cell is where the cellular interrogation takes place, the objective being very simple: to allow the cell of interest to pass through the laser (or lasers) interrogation point and then for the product of that interrogation to be displayed as physical and fluorescent properties of that cell.

The cells must be in a single cell suspension for accurate interrogation. They are put into a test tube then placed on to the sample injection port of the instrument. The sample is delivered to the flow cell by one of two possible mechanisms. The first is to aspirate the sample using a syringe, which then in a dual role delivers the sample to the flow cell by user rate-defined delivery. The second mechanism is perhaps the more common, whereby the user determines one of a series of pressures to be applied within the sealed sample tube, forcing the sample through the flow cell by positive pressure. The cell journey has to be very precise in terms of both speed and trajectory through the flow cell to ensure optimal excitation of any fluorochromes and to gain optimal scatter characteristics from the cells.

This alone however will not ensure that the cells are coincident with the laser beam. There are two mechanisms which need to be active: hydrodynamic focusing and laminar flow. Within the flow cell sheath, fluid is

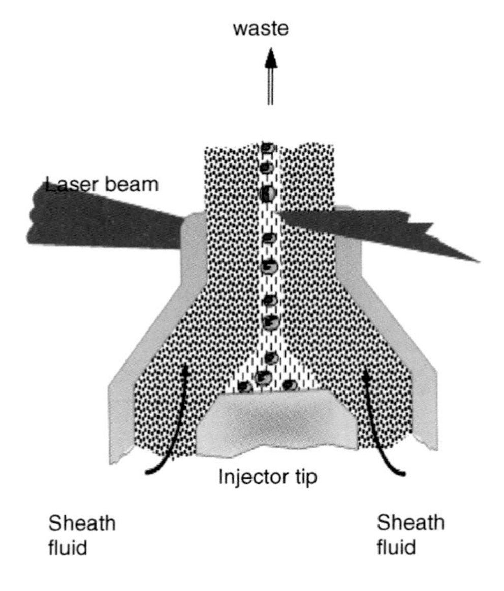

Figure 3.5 A diagrammatical representation of a typical cross-section present in an analysis flow cell, showing the laser interrogation point. Sheath fluid enables hydrodynamic focusing of the cells within the sample stream, which is flowing from the injector tip, past the laser, and then to waste

constantly running at a typical pressure of 4.5–5.0 psi. This acts as a focusing force for the sample core as it is introduced into the flow cell where the sample core is hydrodynamically focused. The cells should then pass the centre of the interrogation point of the laser beam. Sample and sheath fluid will not come into contact because of the laminar flow system that is created, separating the two fluids. Once the cells are analysed they go to waste, along with the sheath fluid (Figure 3.5). Where the sheath-fluid pressure and flow-cell dimensions should be constant, increasing sample pressure will widen the sample core diameter to accommodate the larger volume of sample introduced by increasing the sample pressure (or rate, in systems that are syringe-driven). The sample velocity remains constant.

3.3.2 Laser interrogation at the flow cell

The parameters derived from the cells are forward scatter and side scatter, or large-angle light scatter. Forward scatter is scattered laser light measured

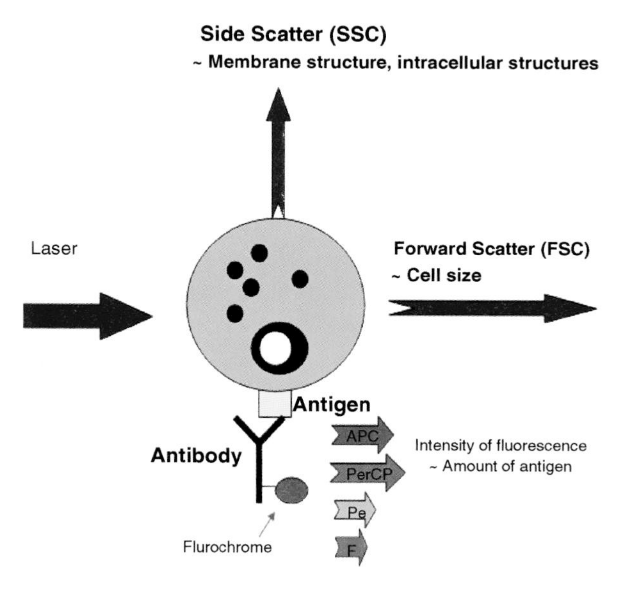

Figure 3.6 Diagrammatic representation of the plan view of an antigen-bearing cell, showing the relative directions of laser, forward-scattered light, side-scattered light and incident direction of the laser beam

in the line of the laser, where it is detected at an angle of <10° and is representative of an approximation of size. Side scatter measures laser scatter and emitted fluorescence at 90° to the laser and is representative of an approximation of intracellular granularity and surface topography. The diagrammatic representation of the flow cell in Figure 3.6 shows both hydrodynamic focusing and laminar flow, each contributing to the correct interrogation of the cell by the laser beam.

When the laser interrogates the cell, the parameters derived are forward and side scatter (both from the 488 nm laser scatter). The forward and side scatter can give a lot of information with regard to the physical attributes of the cell, but most analysis is done by using the flow cytometer to measure fluorescent probes. The number of fluorescent probes detectable by a single instrument has escalated to 18, and by the time you are reading this, probably more, enabling flow cytometry to produce vast amounts of data on relatively small numbers of cells.

Carefully selected fluorochromes can be used simultaneously, but for each the excitation and/or emission spectra must be distinct. An aid to expanding the choice of fluorochromes is the ability to insert multiple lasers into flow cytometers, enabling more excitation lines.

3.3.3 The rate of analysis

Dependent upon the flow cytometer, the rate of analysis (i.e. the speed at which the cells can be accurately interrogated) will vary from 3000 events per second (older type instrumentation) to 20 000 events per second and above for the new 'digital' instruments. These limits are important for the generation of good data representation of the cell population under analysis. Exceeding the limits of an instrument's electronic circuitry will lead to electronically-aborted events that it cannot measure accurately due to the excessive sample throughput rate (Figure 3.7). The data-rate limit therefore is a reflection of how quickly the instrument can see a cell, measure the parameters and reset itself to analyse the next cell. This is commonly known as the 'dead time' of the instrument. These dead time values will reflect how quickly events can be processed by each individual instrument and the effect of varying dead time limits can be seen in terms of beads of varying known concentrations analysed by instruments of differing acquisition capabilities split into analogue and digital instrumentation groups.

The measurements are enabled by laser interrogation of the cells from within the quartz flow cell, and the analysis by a process called pulse processing of the representative photons released. The photons are converted to a proportional voltage (V) by use of a photomultiplier tube (PMT). The PMT detects photons that are reflected towards it and produces a proportional number of electrons by the use of a photocathode.

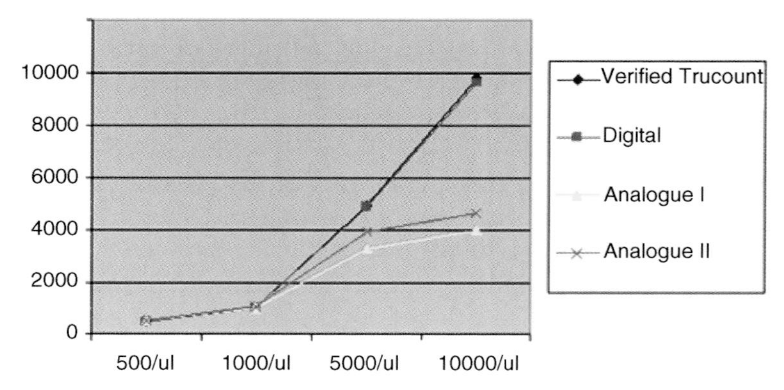

Figure 3.7 A comparison between counting methodologies. *x*-axis shows concentration of 5 micron beads. *y*-axis shows that counts achieved for digital instruments are close to theoretical values (verified Trucount), but for analogue instruments after 3000 events per second errors are encountered due to slower signal processing and consequent aborting of events

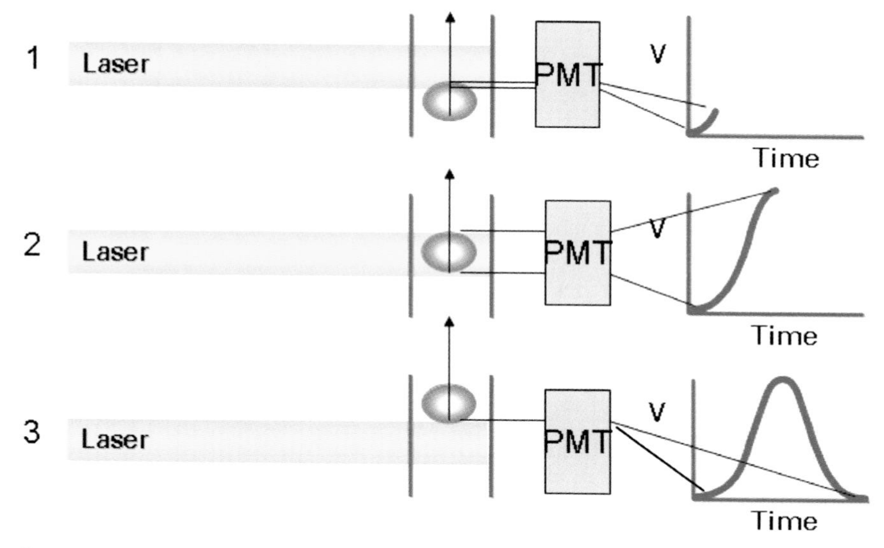

Figure 3.8 A single cell is travelling through the interrogation point of the laser in steps 1 to 3, creating the appropriate pulse with respect to time

These electrons can be 'multiplied' (or not) by the operator to achieve the sensitivity required for a particular experiment; this is usually a function of negative cell autofluorescence. The result is the measurement of a photon-proportional voltage and the creation of a pulse over time of a single cell transit in the laser beam (Figure 3.8).

3.3.4 Pulse processing

A pulse is generated and is representative of each measured characteristic of the cell under interrogation; this then has to be displayed as meaningful data. This is done by measuring either the pulse area or the height. Both these parameters are usually preset by the instrument manufacturer, although on more modern instruments pulse height, area, width and ratio of signals are available for each parameter and can be selected by the user.

The voltage measured after the PMT is usually a maximum of 10 V, so a cell with no fluorescence = 0 V, a cell with maximum fluorescence = 10 V and the cells that are in between will fill the 0–10 V gap to form a fluorescent-intensity histogram. The filling of the histogram is dependent upon the power of the analogue-to-digital co-processor (ADC) and the relative fluorescent intensity of the cell.

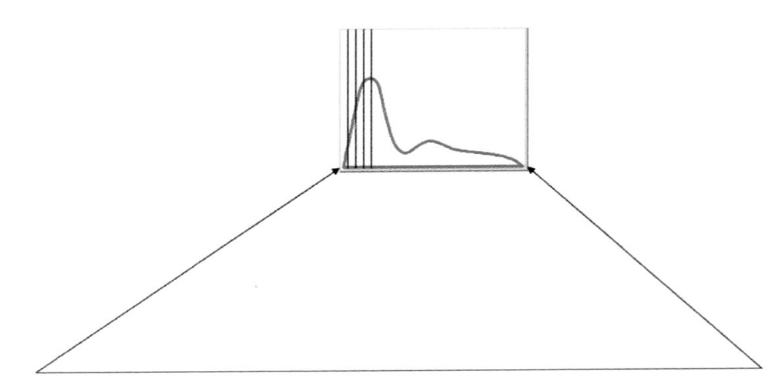

Figure 3.9 The distribution by frequency of cell number (*y*-axis) and intensity of signal (*x*-axis). 1024 channels; each corresponds to a voltage generated by photons emitted either from the laser beam directly or as secondary emission from the fluorochrome. Voltage is usually 0.009 V per channel, so 10 V = 1024 channels

An ADC converts the analogue voltage from the PMT to a digital value proportional to the initial value. Most standard bench-top instruments will have an ADC that is 10 bits (2^{10}), which allows the pulse to be measured and distributed within 1024 bins, or channels, in the histogram. Instruments now coming on to the market have an output of 2^{18} and are able to distribute signals over 262 144 channels, and hence increase the resolution of data (Figure 3.9).

There is now one more decision that needs to be taken regarding the visualization of the cell data, and that is whether to distribute the data in a linear or a logarithmic fashion. This decision is based on the degree of variation shown by the cell parameter under analysis, for example phenotyping of lymphocytes where CD4 positives can be 100 times as bright as the negatives, or in DNA analysis where the G0 population is half the fluorescence value of the G2/M dividing sample. The data that is collected is all initially linear where the measured analogue voltage is from 0–10 V linear scales, as described previously. If, however, the cells are very dissimilar in relative fluorescent intensity then either the signals must go through a logarithmic amplifier to compress the bright and amplify the dim data, so enabling the whole data to be displayed, or the channel numbers within the histogram must be distributed in a logarithmic or a linear fashion (this however relies upon a high number of channels making up the histogram, as in a 2^{18} ADC system; Figure 3.10)

3.3.5 Sample handling by the flow cytometer

Consider the example of a cell with four fluorochromes attached to it.

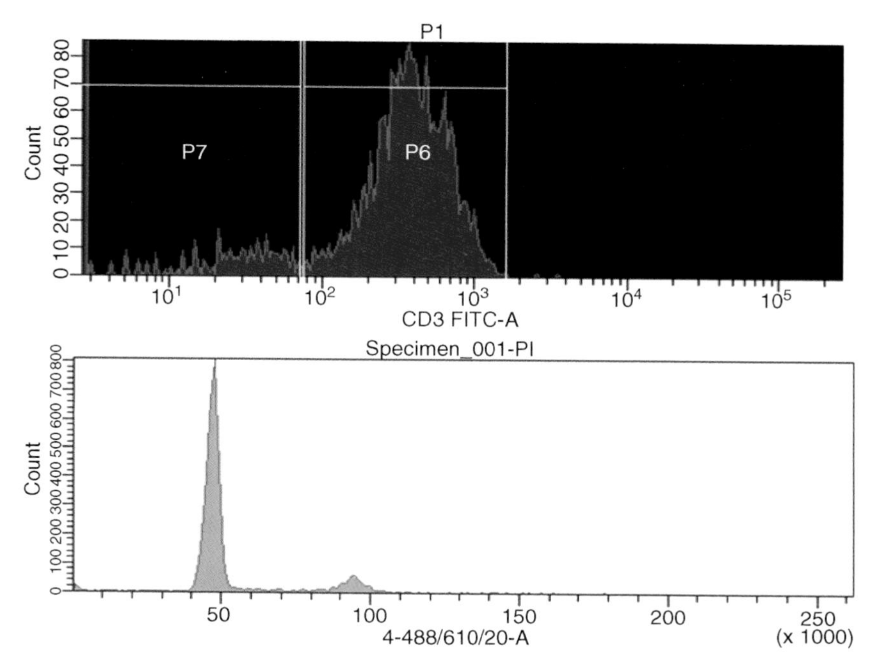

Figure 3.10 Distributing the data in logarithmic and linear plots. The logarithmic distribution of CD3-negative (P6) and CD3-positive (P7) cells. Positive cells give up to 100 times brighter fluorescence than negatives. The lower plot shows a typical cell-cycle analysis, where the initial G0 population is placed on to the linear scale so as to visualize the G2/M population with approximately twice the fluorescent value of the G0

The signal that is emitted from the cell will be forward-scattered light (488 nm laser light), side-scattered light (488 nm laser light) and any fluorescence signals, for example FITC (525 nm), PE (575 nm), PerCypCy5.5 (680 nm) or APC (660 nm). Several other fluorochromes could be used, provided the flow cytometer is equipped with the appropriate fluorescent detectors. This now brings us to a very important aspect of flow cytometry, namely which fluorochromes to pick and how to set up the flow cytometer to detect the fluorochromes of interest. Before we consider the use of the most common fluorochrome, FITC, let us recap on some basic facts defining what a fluorochrome is and why it fluoresces.

A fluorochrome absorbs light from an excitation source, in this instance the laser beam, which results in the electrons of the fluorochrome entering the next electron orbital; an unstable state. For the molecule to then reach a stable conformation, the absorbed energy is dissipated by heat and vibration, resulting in the emission of a longer wavelength of photons (the emission wavelength for the molecule). It is important to note that when looking at the specifications for fluorochromes the maximum excitation of

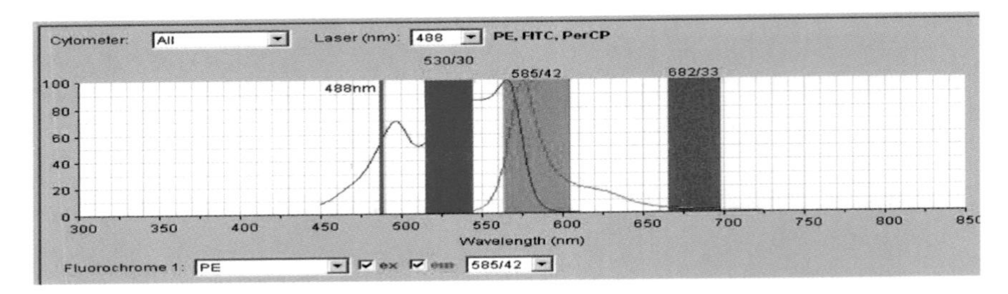

Figure 3.11 Excitation and emission spectra. A typical example of excitation by a blue laser at 488 nm of PE, which has a maximum excitation far to the right. PE is one of the more common fluorochromes used on instruments equipped with blue lasers

the laser required need not be exactly, for example, 488 nm. The reason for this is that excitation of each fluorochrome takes the form of a spectrum, thereby enabling lasers of specific wavelength to excite areas of the excitation spectrum away from the peak, although sub-optimally. Also of note is that, independent of excitation wavelength, the emission of each fluorochrome will be the same. In Figures 3.11 and 3.12, the excitation and emission spectra of phycoerythrin can be seen. Note the excitation maxima of 568 nm and the relative position of the 488 nm laser excitation used to visualize this fluorochrome.

It should also be remembered that different fluorochromes have different 'brightnesses', a function of the extinction coefficient and quantum fluorescence yield of each molecule.

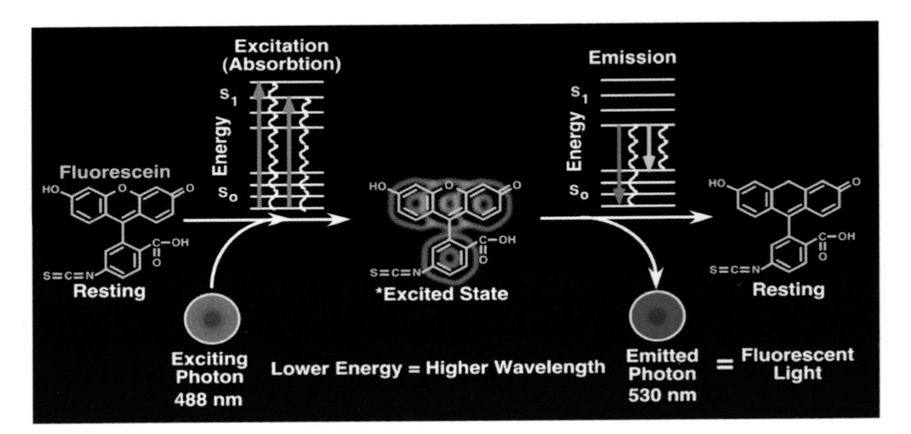

Figure 3.12 The mechanism of blue laser excitation on fluorescein; the electron excitation and resultant fall back to the ground state of the electrons, which produces photons of a longer wavelength and gives rise to the emission wavelength of fluorescein

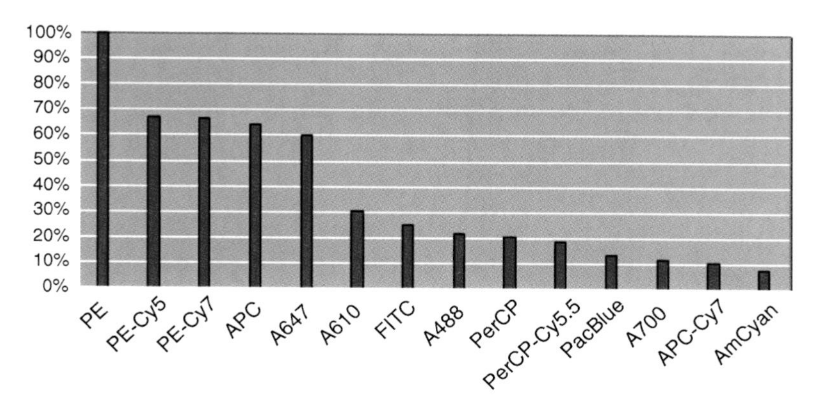

Figure 3.13 Ranking in order of brightness on a Becton Dickinson LSR II of some common fluorochromes

As a general rule, the choice of fluorochrome is usually based on the principle that the weaker the antigen expression, the brighter the fluorochrome that should be used (Figure 3.13).

3.3.6 Sample specificity

If we take as an example the labelling of a CD3 T-cell antibody incubated with a sample of lymphocytes, there are two reactions that take place (Figure 3.14):

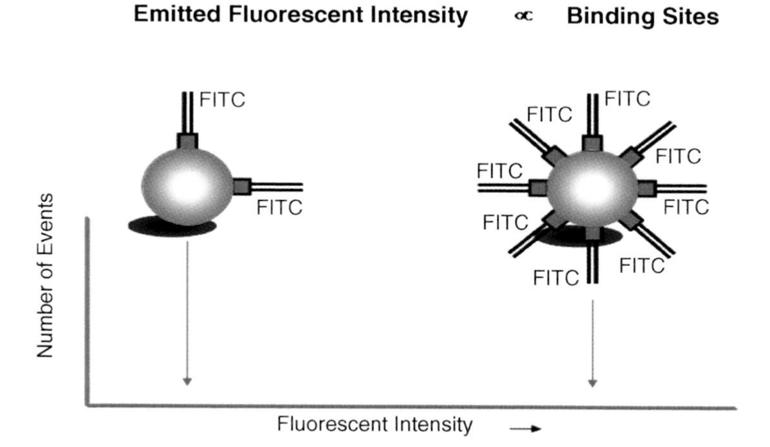

Figure 3.14 Diagrammatic representation of nonspecific and specific binding of an FITC-conjugated antibody where the FITC-conjugated antibodies attributable to specific staining are proportional to the antigen sites on the cell

(1) Specific binding via the interaction of the anti-T-cell antibody with the T-cell antigen, so defining T cells within the lymphocyte population. The intensity of the staining is proportional to the number of antibodies bound on to the T lymphocyte, this being directly proportional to the number of antigenic sites.

(2) Nonspecific binding of the antibody on to non-T cells within the lymphocyte population. This is at a much lower intensity than the specific staining and can be attributable to complement receptor binding and cross-reactivity of the antibody to non-T-cell antigens.

Diagrammatically, the effect of antibody staining can be seen on all cells, not just the target population.

The steps for optimal cytometer set-up for the detection of these respective negative and positive populations are very straightforward.

Defining what is negative is the most important step in flow cytometry. This can be achieved by the use of an isotype control, where cells are incubated with the same protein concentration, fluorochrome and immunoglobulin subtype as the specific test antibody. Occasionally however, isotype controls can have very different staining characteristics to the specific antibody in terms of nonspecific binding, and it may be better to use totally unstained cells or, better still, the negative population of the positively-stained sample, to adjust the settings of the PMTs. This usually places the negative population within the first log decade, provided of course there is a well-defined negative population present.

The photons released from the antigen–antibody-specific and nonspecific FITC-complex are relayed to the PMT allocated for the FITC signal. The photons are steered towards the PMT using a mixture of optical filters, which allow light of the specific FITC wavelength to be delivered and the fluorescence frequency histogram to be produced.

3.3.7 Spectral compensation

It is essential that, in the case where two or more fluorochromes are being used within the same sample, any spectral overlap between the fluorochromes is eliminated, thus ensuring that, for example, the FITC signal represents only the antigens marked by the FITC-conjugated antibody and not those affected by any spectral overlap from a PE-conjugated antibody, and vice versa. Also of importance is the accurate measurement of cells that contain both fluorochromes by virtue of both antigens being present on the cell surface. In Figure 3.15 the spectral overlap for both fluorochromes,

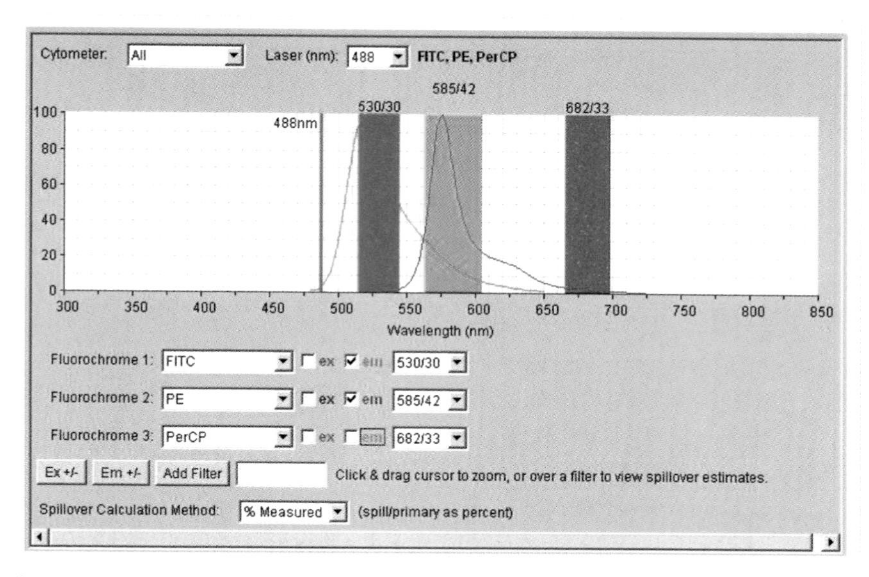

Figure 3.15 Capturing the emission spectra. This diagram shows two band-pass optical filters, shown as 530/30 to capture FITC and 585/42 to capture PE signals, as well as the tail of FITC that goes into the PE measured area, and a small amount of PE that enters into the FITC measured area (spectral overlap)

FITC and PE, can be seen. Also, there is the capacity to see the extent to which the spectra overlap within the respective shaded areas. In addition, the objective positioning of the four major populations to be considered when adjusting compensation using negative-, single- and dual-stained populations can be observed (Figure 3.16).

The resultant dot plot when the FITC (green) signal is allowed to spill into the PE (red) channel can be seen in Figure 3.17.

This is a very important aspect of flow cytometry, where it is common to use up to eight colours in some routine analyses. In order to ensure that the results are accurate and representative of the cell phenotype and/or genotype, compensation has to be implemented with a great deal of care and precision.

3.4 Clinical applications

Now that we know how the histograms and dot plots are generated by the flow cytometer the interpretation is relatively simple. But as I have shown, the data produced is very heavily reliant on good sample preparation and careful operation of the flow cytometer. Major clinical applications where

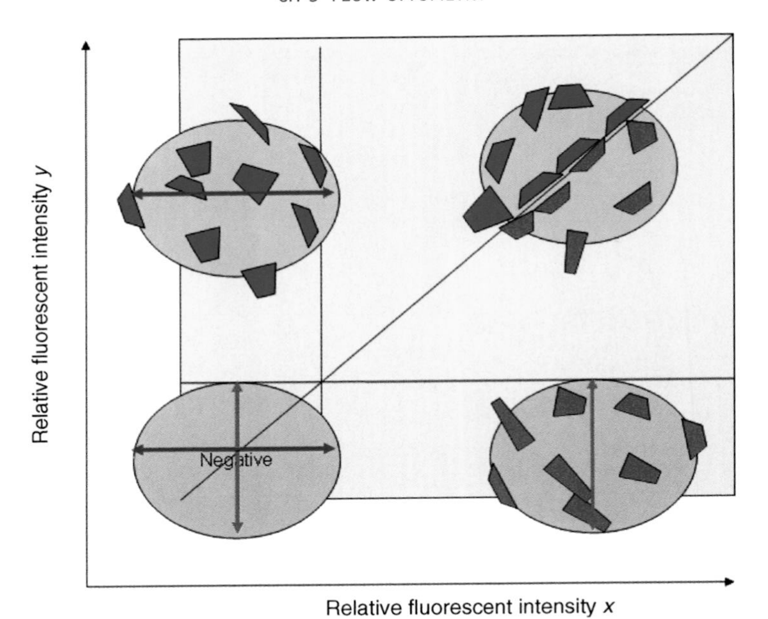

Figure 3.16 Diagrammatic representation of compensation. Single, green-only cells should appear on the *y*-axis; single, red-only cells should appear on the *x*-axis; cells that have equal amounts of green and red should be on the 45° line. In order to place the cluster of cells in the correct position, the red-only cell cluster should have the same *y*-axis-median fluorescent intensity as the negative cells. Also, the green-only stained cells should have the same *x*-median fluorescent intensity as the green-stained cells

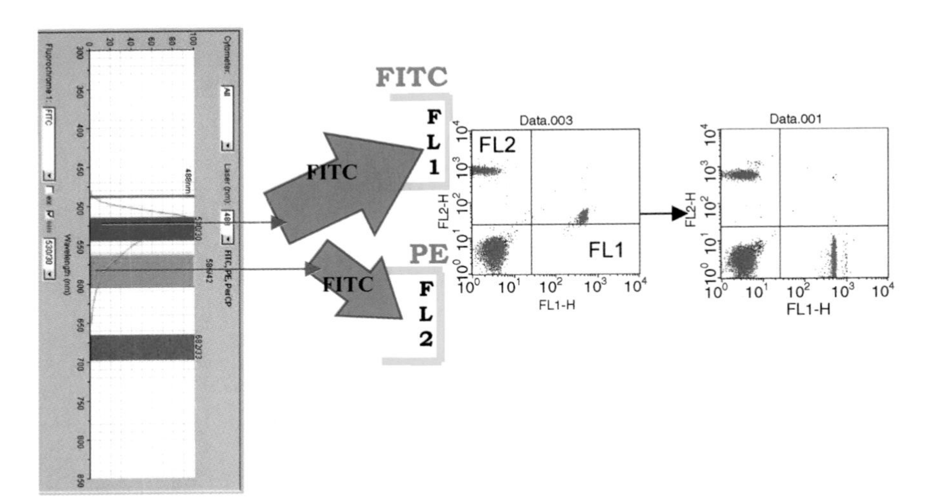

Figure 3.17 Compensating for overlapping spectra. The first dot plot is an example of how a plot will look in uncompensated data and the second dot plot shows correctly compensated data

flow cytometry is used include the diagnosis and subclassification of acute leukaemia and chronic lymphoproliferative disorders, including chronic lymphocytic leukaemia and non-Hodgkin lymphoma, HIV monitoring and DNA analysis.

3.4.1 Haematological malignancies

Flow cytometry acts as an invaluable aid to the diagnosis of haematological malignancies. It is one part of a complex piece of diagnostic pathology that includes morphology, histology, cytochemistry, immunocytochemistry, cytogenetics and molecular genetics. There is inevitably a large panel of lineage-specific markers for each of the cell types that can be involved in the malignant process. Below are some examples of dot plots from pathology samples.

3.4.2 T-lymphoblastic lymphoma

Figure 3.18 shows a malignant T-cell disorder giving rise to the clonal expansion of immature T cells. The young male patient presented with a mediastinal mass and pleural effusion. The total leucocyte count in the pleural fluid was 55×10/l, of which approximately 77% were blasts and 23% small lymphocytes. A few neutrophils, mesothelial cells and macrophages were also present. The dot plots in Figure 3.18 show the reactions with a range of fluorochrome conjugated monoclonal antibodies. The blasts have been gated separately on the basis of their CD45 expression and side scatter and, instead of a standard plot, presented by plotting side scatter against CD45 immunoreactivity. CD45 (leucocyte common antigen) is differentially expressed on all human leucocytes, allowing gating of specific cell populations with subsequent exclusion of cells of no interest to the current experiment. Subsequent plots illustrate the percentage of cells positive with the respective antibody. This analysis was consistent with a diagnosis of T-lymphoblastic lymphoma, with a common thymocyte phenotype (CD3-ve, cytoplasmic CD3+, CD7+, CD2+, CD5+, CD4+, CD8+, TDT−/+−).

Any residual normal haematopoietic elements from Figure 3.18 can also be evaluated, as the example in Figure 3.19 shows.

3.4.3 Acute myeloid leukaemia (AML)

There are varying degrees of maturation and differential granulocyte involvement within the AML group. This example shows minimal

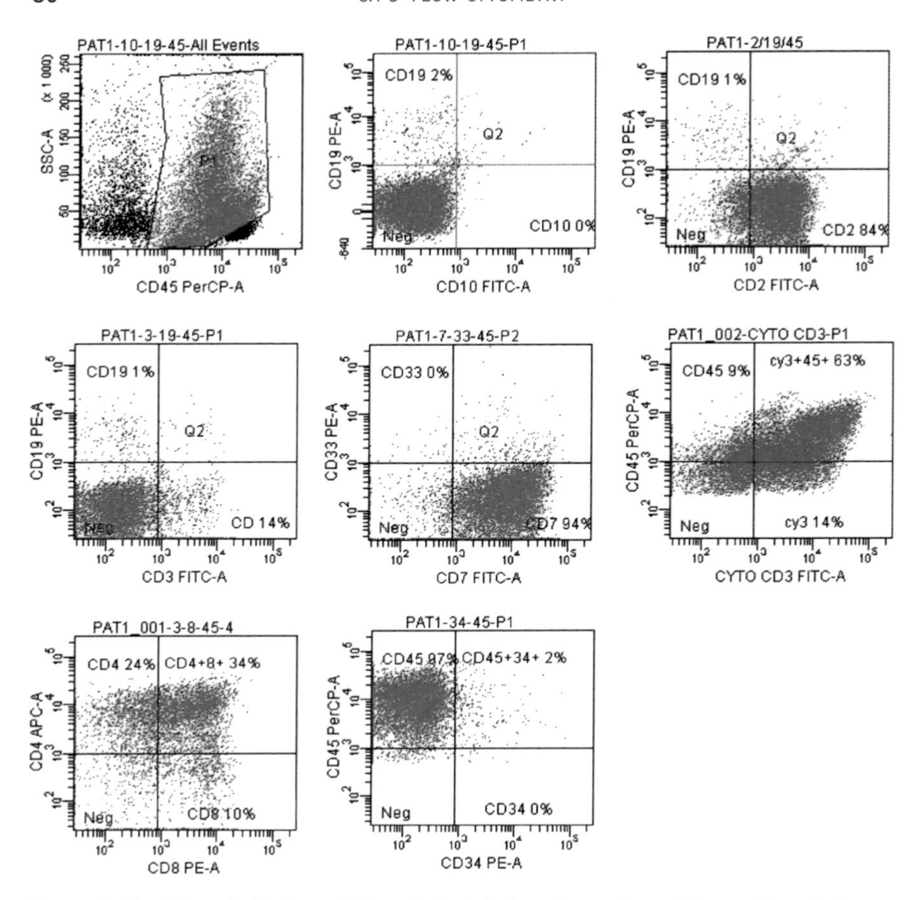

Figure 3.18 Pleural effusion of T-lymphobalstic lymphoma. From left to right, the large lymphoma population is defined using (1) SSC vs. CD45 to give CD45+ and CD45− populations by gating. The CD45+ in the major lymphoma population is displayed in terms of the following antigens: (2) CD10/CD19; (3) CD2/CD19; (4) membrane CD3/CD19; (5) CD7; (6) cytoplasmic CD3/CD45; (7) CD4/CD8 and CD8; (8) CD34/CD45. Note that the use of cytoplasmic markers is very useful in instances where very immature cells may contain a lineage-specific marker intracytoplasmically, such as T-cell specific CD3, but where the cell has not matured sufficiently to demonstrate this marker on its membrane. Also, the gating excludes a small mature lymphocyte population strongly CD45-positive at the base of the gate, which is analysed in Figure 3.19. SSC, side scatter

evidence of maturation, but some markers are useful in defining maturation, including CD11c, myeloperoxidase (MPO) and CD68. This example shows a homogeneous population when forward scatter is plotted against side scatter. The immunophenotype obtained by gating on this population is largely correct. However, when CD45 expression is plotted against side scatter, a separate CD45-strongly-positive, low-side-

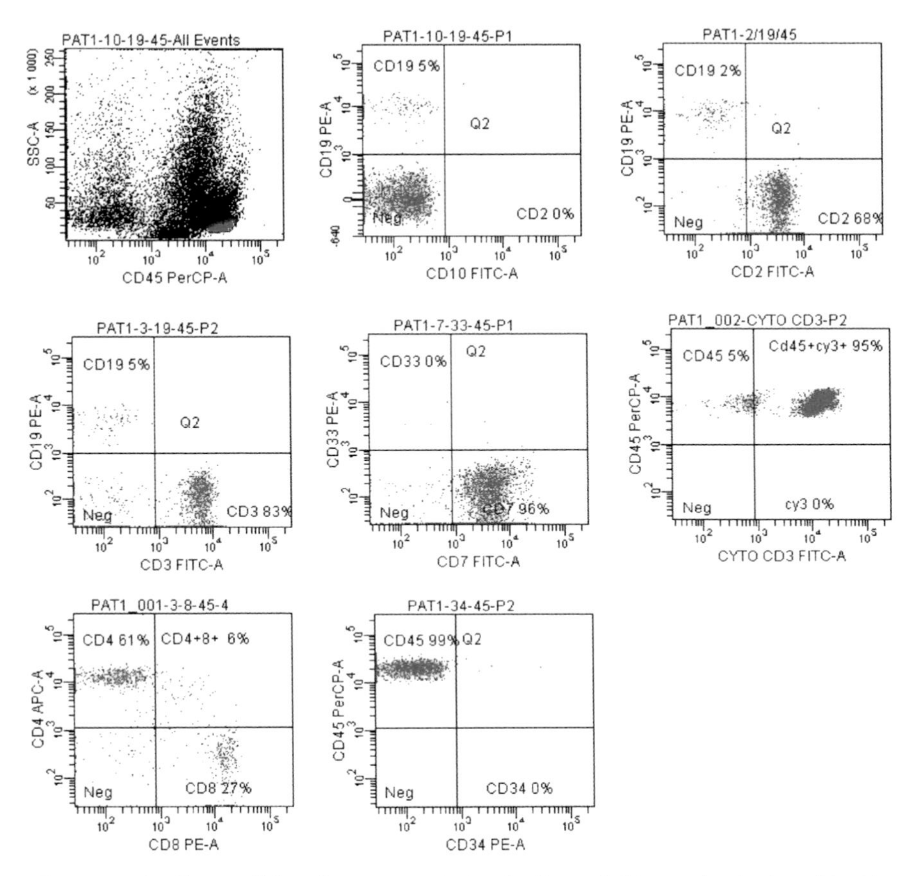

Figure 3.19 The small lymphocyte population in Figure 3.18 can also be typed by the markers used, verifying the mature lymphocyte phenotype. From left to right: (1) CD45 (bright red); (2) CD19/CD10; (3) CD2/CD19; (4) membrane CD3/CD19; (5) CD7/CD33; (6) cytoplasmic CD3/CD45; (7) CD4/CD8; (8) CD45/CD34

scatter population separates from the main concentration of events. This small population represents normal lymphocytes, while the large population identifies the malignant cells. CD45 is often more weakly expressed on malignant cells than on normal counterparts. Thus, a more accurate assessment of the immunophenotype would be obtained by gating out the lymphocytes and analysing antigen expression on the leukaemic cells only. This is good practice and becomes even more essential when the malignant population forms a smaller component of total cells.

Figure 3.20 shows an example of AML with some of the more common phenotypic markers used for flow cytometry.

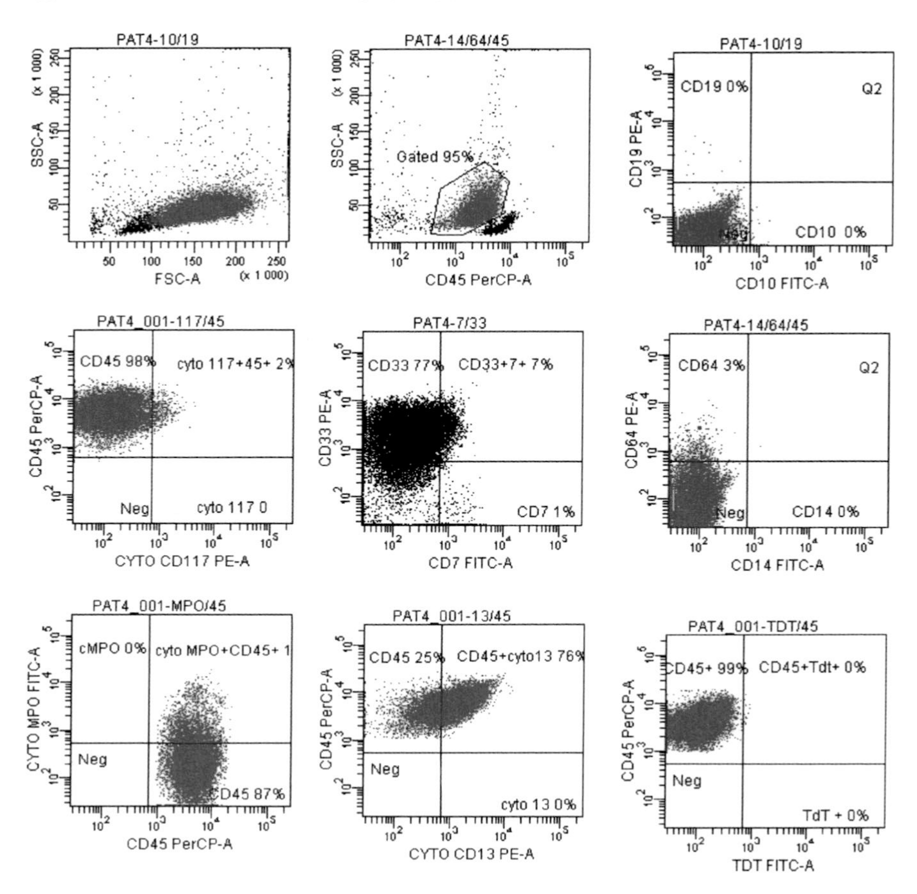

Figure 3.20 Acute myeloid leukaemia (AML). From top-left to bottom-right: (1) scatter, gated, showing a homogenous population of cells; (2) SSC vs. CD45 showing the ungated, lymphocyte population (not coloured) and weaker blast population CD45 (gated, red); (3) CD10/CD19; (4) CD45/cytoplasmic CD117; (5) CD33/CD7; (6) CD14/CD64; (7) CD45 and CD45/cytoplasmic MPO; (8) CD45 and CD45/cytoplasmic CD13; (9) CD45/Tdt. MPO, myeloperoxidase; SSC, side scatter; TdT, terminal deoxytransferase

3.4.4 Human immunodeficiency virus (HIV) monitoring

The monitoring of T- and B-cell subtypes is of particular importance in patients with HIV. In particular, the CD4 count, as opposed to percentage, is monitored very closely with respect to the treatment for these patients. It is also important to discriminate CD4-positive T cells from any other cells that may have CD4 on their surface, such as monocytes that contain relatively lower but still discernible levels of CD4.

The evaluation of percentages by using fluorescent microscopy and monoclonal antibodies to CD4 and CD3 was an early method for this test. Progress to the use of flow cytometry enabled an increase in speed and accuracy. Dual-colour flow cytometry was used in the evaluation of the CD3+ /CD4+ T cells as a percentage, using a gate specific to lymphocyte forward- and side-scatter properties. The next modification was to add a CD45 to make this a three-colour assay, so that the slight inaccuracy and user-variability from the scatter gate was eliminated. The CD45-negative debris was also eliminated from the gate.

The accuracy of results was improved, and instead of reporting percentages of CD4 cells, absolute counts were reported. This gave the benefit of a finite measurement of a patient's CD4 status, as opposed to a relative percentage dependent on the levels of the other lymphocyte elements present. There are two different methodologies for achieving an absolute count on a flow cytometer, which are very much instrument-specific. (1) The use of a flow cytometer that will measure a finite volume during the analysis. This will be from either a syringe mechanism or a direct volumetric measurement (Partec) and will therefore calculate the absolute number of cells from the percentage of cells and volume analysed. (2) The use of beads in a sample calibrator. This is the most common method, whereby the beads of a known concentration are placed within an exact volume of sample. A ratio of beads to cells is observed and the absolute count can be calculated by from this.

The example in Figure 3.21 is from a six-colour HIV pre-mixed antibody cocktail produced by Becton Dickinson. The calibration beads for the count are already distributed into the tube used for the analysis, and the concentration of the beads is entered into the software to enable automatic evaluation of the absolute counts. The operator adds the appropriate sample volume of blood and antibodies to the tube, then incubates and lyses the sample. The sample is placed on the flow cytometer; the instrument settings are optimized by an automatic instrument calibration procedure for this type of assay so that the sample is analysed without any user intervention, unless editing of results is required. The software then calculates the counts and percentages. This type of automated analysis is now necessary for what is a very high-throughput test. Figure 3.21 shows an automatically-generated report form a six-colour HIV assay, employing a protocol using a manufacturer-produced pre-mixed antisera that can be automatically configured to run using calibrated bead-generated PMT and compensation settings.

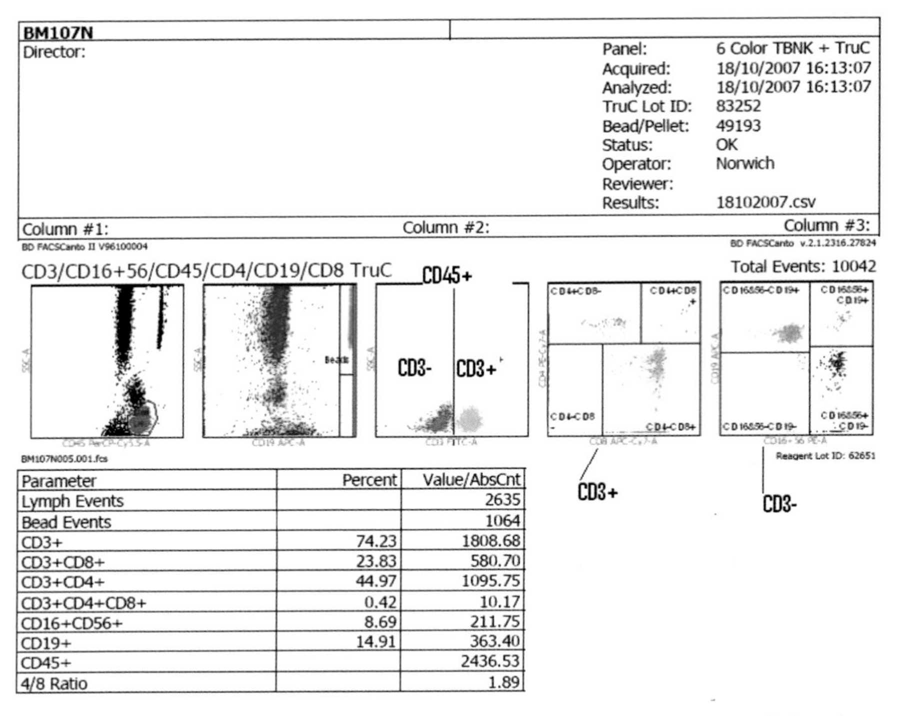
BM107N

Director:

Panel:	6 Color TBNK + TruC
Acquired:	18/10/2007 16:13:07
Analyzed:	18/10/2007 16:13:07
TruC Lot ID:	83252
Bead/Pellet:	49193
Status:	OK
Operator:	Norwich
Reviewer:	
Results:	18102007.csv

Column #1: Column #2: Column #3:

CD3/CD16+56/CD45/CD4/CD19/CD8 TruC

Total Events: 10042

Parameter	Percent	Value/AbsCnt
Lymph Events		2635
Bead Events		1064
CD3+	74.23	1808.68
CD3+CD8+	23.83	580.70
CD3+CD4+	44.97	1095.75
CD3+CD4+CD8+	0.42	10.17
CD16+CD56+	8.69	211.75
CD19+	14.91	363.40
CD45+		2436.53
4/8 Ratio		1.89

Figure 3.21 Human immunodeficiency virus (HIV). This shows an automated six-colour HIV screen. From left to right: (1) SSC vs. CD45; (2) SSC vs. CD19 including calibrant beads for absolute counting; (3) SSC vs. CD3 and CD3; (4) CD4 vs. CD8 showing at the top left; this plot is gated on CD45-positive and CD3-positive; (5) CD19 vs. CD16; this plot is gated on CD45-positive and CD3-negative, depicting CD19- and CD56/CD16-positive cells to make up a comprehensive HIV profile. SSC, side scatter

3.4.5 DNA analysis

The DNA content (ploidy level) and proliferation rate of the cells can be determined by the proportion of cells within each discernible phase of the cell cycle. Cells with 2N DNA can be classified into G0 (non-cycling)/G1 (pre-synthesis growth, G0G1), S (DNA replication containing between 2N and 4N amounts of DNA) and G2 (post-synthesis growth)/M (mitosis).

If a single parameter evaluation of DNA content is carried out, the compartments identified are G0/G1+S +G2/M. The most common stain for DNA analysis is propidium iodide (PI). This, upon permeabilization of the cell, will bind one molecule to approximately five base pairs of DNA. It will also bind to RNA, so RNase treatment is necessary prior to staining. Permeabilization is a very important step in the detection of DNA within cells by dyes that do not have the ability to pass through the membrane. There is a necessity with dyes such as PI to 'puncture' the cell gently in order to allow pores to develop in the cell membrane, thus allowing the DNA to be exposed to chelating dyes. The most common way to do this is to place the cells drop-wise into 70% ethanol at 20 °C. The stoichometric association of PI to DNA means that differential levels of DNA and therefore the different compartments of the cell cycle can be detected within the sample. The G2/M population occupies a position on this linear histogram at twice the channel number of that of the G0/G1 population, representing the increase by a factor of two of the DNA content present within the G2/M phase cell. One possibility of error in determining an accurate G2/M population is that if two G0/G1 cells are clumped together, the flow cytometer may see and place these (2N DNA) in the position of one G2/M cell (4N DNA). To minimize the risk of this happening, electronic circuitry called a 'doublet discriminator' is in place to interrogate the shape of the signal produced and allow the detection of most of these doublets.

There are also events that can be detected below the G0/G1 population. These can be attributed to either a hypo-diploid population where total DNA content has been lost due to chromosomal deletions or the presence of apoptotic events [1].

Figures 3.22 and 3.23 give some examples of DNA analysis using both PI and 4,6-diamidino-2-phenylindole (DAPI), showing cycling cell profiles.

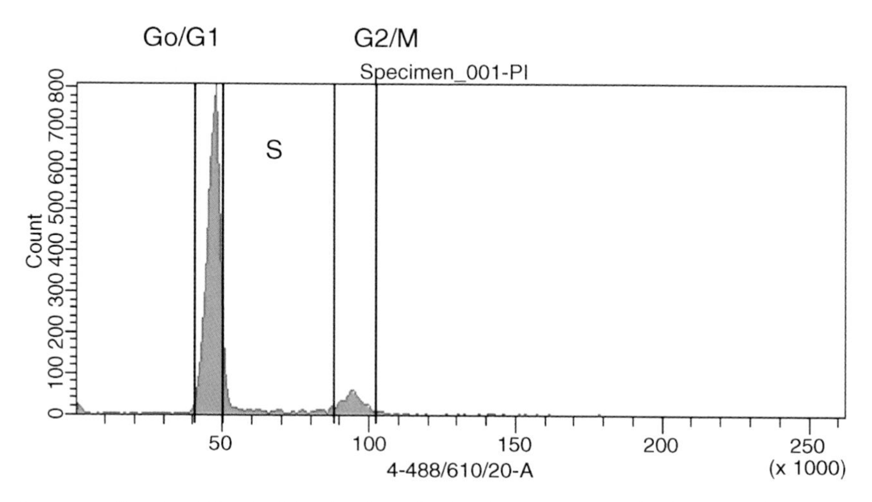

Figure 3.22 Propidium iodide staining of cells after treatment with RNase and excitation with a blue laser. Data has been gated on a doublet discriminator

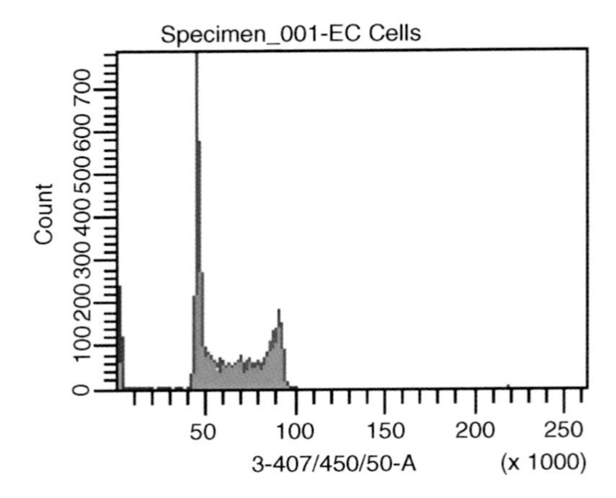

Figure 3.23 A proliferative embryonic carcinoma cell culture. Excitation of DAPI staining by a violet laser. DAPI, 4,6-diamidino-2-phenylindole

One of the most standard ways of detecting sub-G0/G1 cell-cycle apoptosis is to detect the peak to the left of the G0/G1 peak. However, there are more accurate methods.

Figure 3.24 gives an example of manipulating the different cell-signal profiles (area and width in this example; area and height can also be used) generated from the laser interrogation of the cells to help exclude doublets from being included in the DNA profile and so giving erroneous results.

There are other dyes that allow you to avoid permeabilization by virtue of the fact that they will cross the intact cell membrane. The most popular of these is Hoechst. This dye has the added advantage of not staining RNA, so RNase treatment is not required. It is selective for A-T regions of DNA and binds within the minor groove. DAPI also associates with the minor groove of DNA, again binding to A-T clusters; however, it will not cross the

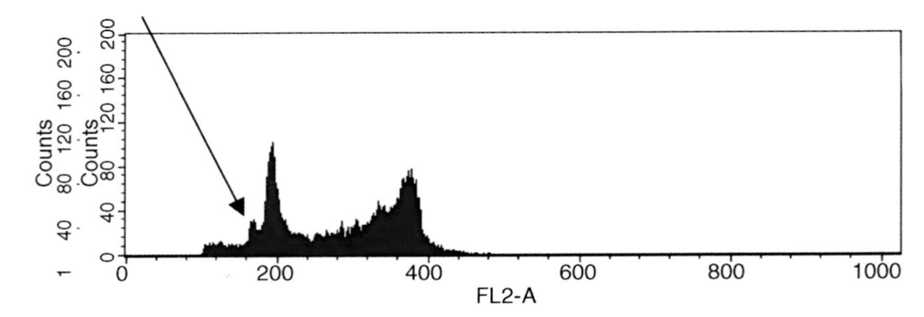

Figure 3.24 Area and width manipulation to show a small apoptotic sub-G0/G1 peak

membrane of the cell, so permeabilization is a required step if using this dye. Relatively new dyes such as DRAQ5 (Biostatus) and the Vybrant range (Invitrogen) will also permeate the cell membrane to stain DNA.

The use of standard histogram statistics as supplied by most flow cytometer companies will give a statistical evaluation on manually-gated and user-determined cell-cycle phases. However, it can be argued that a more accurate way of doing this is to curve-fit with software such as that supplied with FlowJo or 'ModFit'. The advantage of such software is that there is an extrapolation by standard algorithms for G0/G1, S, and G2/M populations of the cell cycle, so a more accurate estimation of the cell-cycle components can be achieved (Figure 3.25).

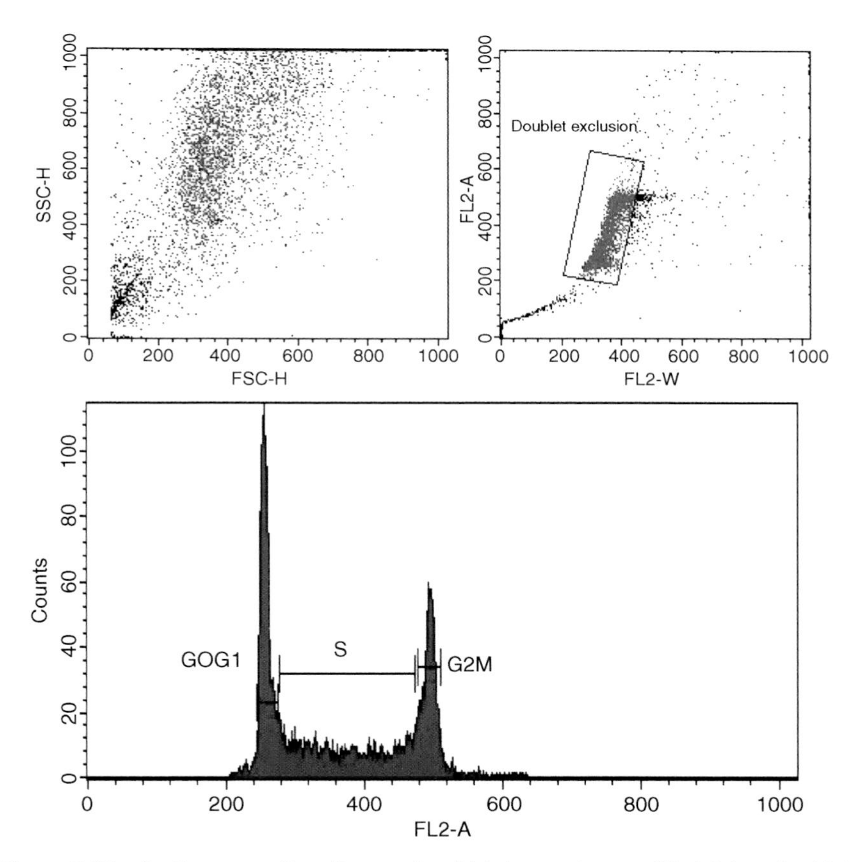

Figure 3.25 Scatter properties of a sample which has undergone PI staining. Doublets within the sample are then excluded using the signal-processing and gating capabilities of the area and width signal associated with the PI staining. Finally the gated data is represented in terms of the area signal output in order to evaluate each cell-cycle compartment

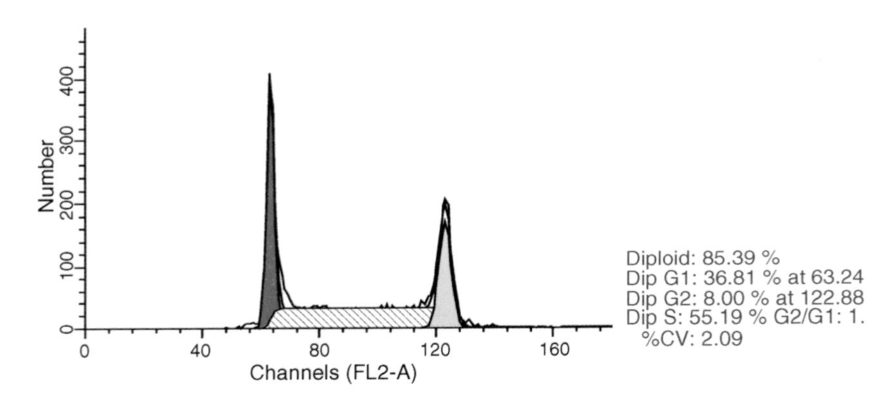

Figure 3.26 Accurate assessment of DNA compartments. A curve-fitting software program (ModFit) is used to more accurately fit the components of the DNA profile

In Figure 3.26 the different compartments of DNA are shown, but in order to calculate these accurately it is usual to use curve-fitting software. This is done so as to separate cells in early S phase from G0/G1, and cells in late S phase from those in G2/M.

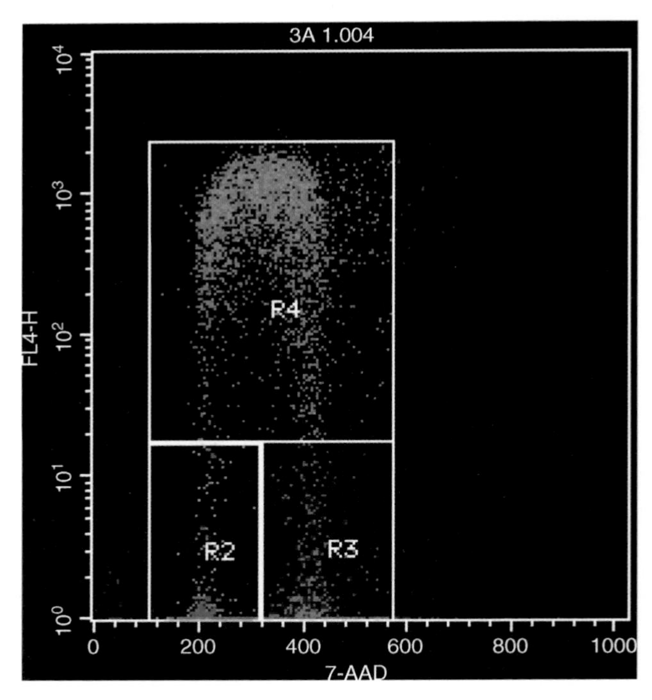

Figure 3.27 The use of 5-bromo-2-deoxyuridine–allophycocyanin (BrdU-APC) conjugated antibody for DNA analysis. The green plot shows G0 cell cycle with minimal incorporation of BrdU; the blue plot shows maximal incorporation of BrdU in S phase; the red plot shows G2 phase with minimal incorporation of BrdU

DNA analysis is not only achievable on free cells; there are many techniques whereby paraffin sections can be used to look at the DNA within the tissue section [2].

It is useful however to try to see the cycling cells within the G0 population that cannot normally be seen when using a single DNA dye. A technique that employs 5-bromo-2-deoxyuridine (BrdU) can be used. This technique will separate DNA synthesizing cells from resting cells.

BrdU is a thymidine analogue and will be incorporated into the DNA of cycling cells *in vivo* or *in vitro*. After incubation, the cells can be permeabilized and a conjugated anti-BrdU antibody added to allow for the evaluation of the cycling cells, as seen in Figure 3.27.

3.5 The future for flow cytometry

It would appear that the future has caught up with us. We now have flow cytometers that are capable of 20 parameters and can be customized in terms of different lasers and emission optics to suit virtually all clinical and research needs. One thing that is lacking in a conventional flow cytometer however is an assessment of the morphology of cells.

One very interesting instrument that has attempted to put this to rights is the Amnis ImageStream system. The ImageStream uses a fluidic system similar to that of a conventional flow cytometer, but combined with CCD camera technology, optical filtration and digital computing, to image and analyse each individual cell. For each cell, six images are produced simultaneously, including a transmitted light image (brightfield) for morphology, a laser side-scatter image (darkfield) for granularity, and multiple colours of fluorescence, which provide intensity, distribution, co-localization and more than 200 other parameters.

These parameters can be used to generate the dot plots and histograms of standard flow cytometry and allow the scientist to click on a specific dot to inspect the associated cell, view all the cells within a gate drawn on the plot, or back-gate cell images of interest on the plots. The data-analysis software is also able to take user-specified cell populations and automatically determine which parameters distinguish them. The instrument can use up to three lasers and offers the option of extended depth of field so the entire cell can be kept in focus for applications such as fluorescent *in situ* hybridization (FISH) spot counting.

There is a trade-off however, in that with a conventional flow cytometer data rates can be 20 000 events per second, whereas the current-generation

ImageStream will acquire data at only up to 300 events per second. It has to be said that for each cell, the ImageStream will capture six images from two scatter and four fluorescence parameters, thereby yielding 60 000 images per 10 000 cells.

Software is used to interrogate the images for quantitative morphology, fluorescence signal and signal localization (Figure 3.28). The constraints on higher speed are largely financial rather than technological and have to do mainly with the speed of computing. Future versions of the ImageStream should be able to match or exceed the speed of conventional flow cytometers thanks to the increase in desktop computing power, which makes this type of instrument a much-awaited addition to the repertoire of any flow-cytometry core facility.

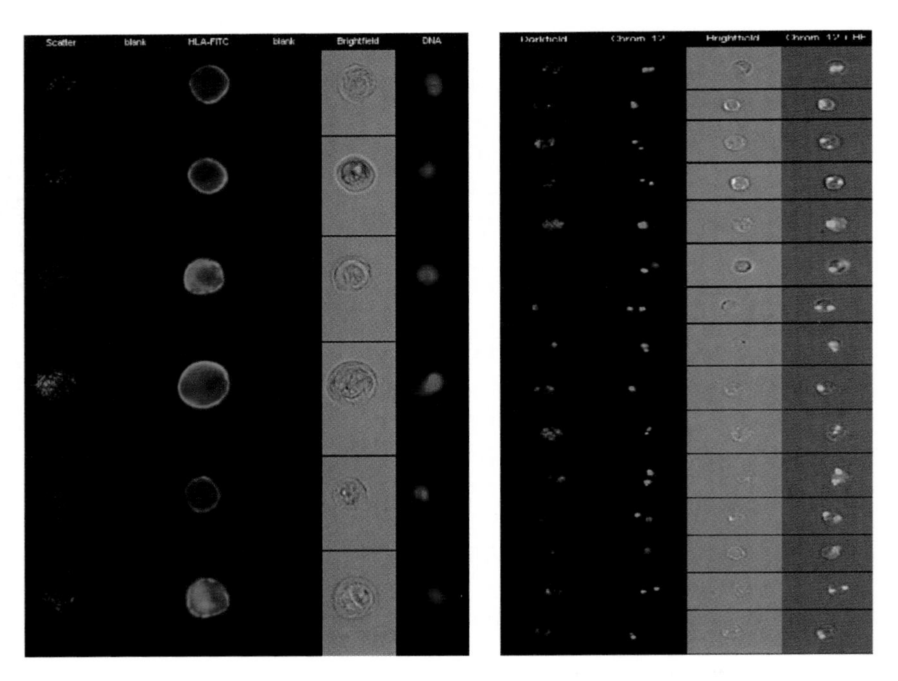

Figure 3.28 Morphology and stain localization data collected from the Amnis ImageStream. Two data sets are shown in the figure. On the left are six mammary epithelial cells, one cell per row, imaged in darkfield (blue), HLA-FITC fluorescence (green), brightfield (grey) and 7-AAD fluorescence (red) to reveal nuclear morphology and ploidy. On the right are human peripheral blood mononuclear cells probed in suspension with a FISH probe against chromosome 12. The cell images include darkfield (blue), chromosome 12-FITC fluorescence (green), brightfield (grey) and a brightfield–FISH probe overlay. FITC, fluoroisothyocyanate; 7-AAD, 7-Amino-actinomycin D; FISH, fluorescent *in situ* hybridization

3.6 Conclusions

The information within this chapter has hopefully given a brief overview of the major features to be considered when operating or analysing data from a flow cytometer. Advances in both flow cytometry reagents (more stable tandem-dye conjugates and the use of nanoparticles with reduced spectral overlap as fluorochromes) and instrumentation are a permanent feature to this field of science. The challenge is not only to keep up to date with the advances so as to deliver the best, most accurate scientific data, but also to build these advances on a very solid basic understanding of the flow cytometer being used and the requirements of the cells or analyte under investigation.

Acknowledgements

I would like to acknowledge the following for the courtesy of using their figures in this chapter:

Becton Dickinson, UK; Figures 3.3, 3.5, 3.6, 3.8, 3.11, 3.13, 3.14 and 3.15.

Prof. Denis Alexander, Belfast Health & Social Services Trust, UK; Figures 3.18, 3.19 and 3.20.

Allyson Tyler, Norfolk & Norwich University Hospital, UK; Figure 3.21.

Mr O Hughes and Dr M Lako, North East England Stem Cell Institute, Newcastle, UK; Figure 3.26.

David A Basiji PhD, President & CEA, Amnis Corporation, USA: Figure 3.28.

Useful Web sites

BD Biosciences: http://www.bdbiosciences.com/immunocytometry_systems/support/training/online/ITF/start.html

Invitrogen: http://probes.invitrogen.com/resources/spectraviewer/ and http://probes.invitrogen.com/resources/

The Scripps Research Institute: http://facs.scripps.edu/.

References

[1] Illera VA, Perandones CE, Stunz LL, Mower DA, Ashman RE. Apoptosis in splenic B lymphocytes. J. Immunol. 1993;151:2965–2973.

[2] Corver WE, Ter Haar NT, Dreef EJ, Miranda NF, Prins FA, Jordanova ES, Cornelisse CJ, Fleuren GJ. High-resolution multi-parameter DNA flow cytometry enables detection of tumour and stromal cell subpopulations in paraffin-embedded tissues. J. Pathol. 2005;206:233–241.

4

Immunocytochemistry

Perry Maxwell[1] and Merdol Ibrahim[2]

[1]*Principal Clinical Scientist, Belfast Health & Social Care Trust, Centre for Cancer Research & Cell Biology*
[2]*Manager, UKNEQAS ICC & ISH*

4.1 Introduction

The distribution of proteins within cells and tissues can be identified using immunocytochemistry (ICC)/immunohistochemistry (IHC) – where both terms are used synonymously. Since its introduction in the 1940s, ICC has become an essential tool in both research and clinical laboratories for the detection of changes in cell proteins. This technique is invaluable for identifying disease and predicting response to therapy, and is used considerably in fundamental scientific research.

It has the advantage over other protein assays that the target protein can be visualized *in situ* either as an intra- or an inter-cellular component depending upon the target protein. Following on from the genomic revolution, where protein-encoding genes were identified, the proteins themselves have returned to centre stage as targets in the fight against cancer and in the battle against infectious diseases.

The clinical applications of ICC have developed in recent years and together with improved methodologies and techniques have become more specific and sensitive in their application and more cost-effective. The introduction of automated ICC has facilitated high-volume testing in the clinical laboratory and complimented high-throughput automated gene discovery in research laboratories. Alongside these technical developments have been improvements in standardization both at the bench and in the reporting of results. Quality has improved with an increased awareness

Advanced Techniques in Diagnostic Cellular Pathology Edited by Mary Hannon–Fletcher and Perry Maxwell

amongst both practitioners and the general public of the impact results have on patient care.

In this chapter we seek to explain these developments and highlight the issues of standardization and quality, with some suggestions as to how these might be achieved. We highlight the work of the external quality-assessment programme based in London, the United Kingdom National External Quality Assurance Scheme for Immunocytochemistry and *In Situ* Hybridization (UK NEQAS ICC & ISH), with examples of results pooled from the 400 + laboratories participating from across the world.

4.2 Basic principles

4.2.1 Fixation and pretreatment

The development of ICC as a diagnostic tool following the introduction of Diaminobenzidine (DAB) as a chromogen in the 1960s remained problematic, due to alterations, cross-linking and masking of proteins introduced during the fixation process. The widespread use of formalin as a primary fixative means that protein–formaldehyde interactions are central to the majority of techniques in cellular pathology. The introduction of proteolytic enzymes overcame some of the problems associated with fixation and allowed the technique to develop with consistent and reproducible results. Some antigens however still remained inaccessible to their antibody following formalin fixation. It was thought that certain monoclonal antibodies were not suited to paraffin wax-embedded material or that detection systems lacked sensitivity. The introduction of microwave-based heat-mediated antigen retrieval [1], initially in metal-containing buffers or urea and later in citrate buffers or EDTA, showed that the problem may have been due to masking of epitopes induced by formalin fixation. It was also shown that using high temperatures in an autoclave or pressure cooker performed the same function as the microwave oven. Some authors advocated substituting the word 'epitope' for 'antigen', thus giving HMER, heat-mediated epitope retrieval.

The papers by Morgan *et al.* [2, 3] showed that the chemistry involved during fixation is quite complex. They proposed that formalin fixation involves the formation of direct cross-linkages with protein side groups, the methylation of amino groups on aromatic amino acids with subsequent formation of methylene bridges, and the time-dependant formation of calcium ordinate bonds. As the optimal demonstration of antigens requires their exposure following the fixation process, studies have focused on methods to overcome these issues.

Morgan *et al.* [3] looked at the effect HMER has on the demonstration of the proliferating antigen Ki67 following formalin fixation. They found the hydrogen ion concentration of the buffer to be a key factor in the mechanism by which epitope retrieval works. It was found that at low pH, there was a reversal of formalin cross-linkages by acid hydrolysis, but with consequent loss of tissue morphology. At high pH, selective chelation of the divalent metal ions from the complexes allowed antibody–antigen interaction, with exogenous calcium inhibiting the retrieval of antigens. There was an assumption that calcium formed ordinate bonds but this has been challenged by Yamashita & Okada [4]; using five proteins fixed in 4% formaldehyde with and without 25 mM calcium chloride, heated at a pH 3.0, 6.0 and 9.0 and analysed on SDS-PAGE, they found that the main mechanism of antigen retrieval was due to disruption of the pH-dependent cross-links.

It has also been advocated that conformational changes of the secondary and tertiary structures of native proteins are important factors in HMER. The retrieval of these structures is dependent on salt concentration and pH [5, 6].

In summary, although the exact effects of formalin fixation on intra- and inter-molecular cross-links have yet to be resolved they do appear to be nullified by HMER. Furthermore, electrostatic and hydrophobic effects are also reversed by HMER, the process of which is pH-dependent.

It should be noted that a false-positive reaction following HMER can be introduced when a biotinylated detection system is used, due to the exposure of endogenous proteins. When used on cytological material, HMER can destroy the morphology of the cells and lead to false-positive reactivity. Care should therefore be exercised in the interpretation of ICC, especially where some form of heat retrieval has been used.

4.2.2 Detection systems

The relative sensitivities of the various antigen-detection methods are well known. Early reports suggested that if an arbitrary value of 1 was assigned to directly conjugated methods then the peroxidase-antiperoxidase (PAP) and streptavidin ABC methods could be assigned as being 100 and 1000 + times more sensitive, respectively. It was proposed in the 1980s that polymers containing multiple copies of label could also be used to amplify the signal and consequently increase sensitivity. Double-stage detection systems, which conjugate a polymer composed of multiple copies of peroxidise, are now readily available and have become the method of choice. Even though they are essentially indirectly-labelled conjugates,

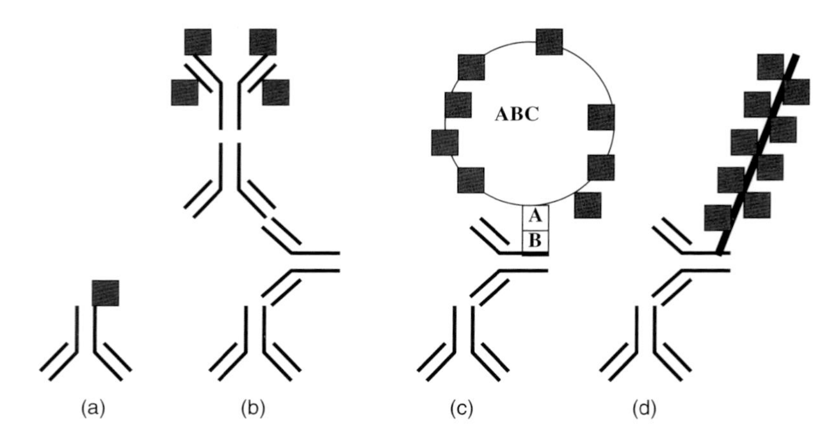

Figure 4.1 Comparison of methods: (a) direct; (b) peroxidase-antiperoxidase (PAP); (c) avidin biotin (ABC); (d) enzyme-labelled polymer method. Black boxes represent the enzyme; A, avidin; B, biotin

their multiple copies of peroxidase mean there is an immediate amplification of signal equivalent to or even superior to strepatvidin ABC methods; especially useful in biotin-rich tissues. Moreover, these alternatives to the standard three-stage methodologies offer a time-saving convenience (Figure 4.1).

Amplification of the chromogen is also enhanced by the deposition of metals such as copper or nickel on to DAB.

4.2.3 Introducing antibodies into the laboratory

The introduction of antibodies to the laboratory needs to be controlled and correctly validated. Most often, the main source of information is the manufacturer's data sheet accompanying the product. This is the starting point. Ideally, the manufacturer should supply details such as the clone, immunoglobulin class and concentration; optimum handling procedures including a recommended dilution as a guide; the method by which such information has been acquired; the distribution of the antigen in normal and abnormal cells/tissue; and a list of references. Each laboratory should then confirm that the guidance offered is appropriate to its fixation and processing schedule. It is recommended therefore that laboratories keep stores of control tissue/blocks that are suitable for the testing of any new antibodies.

For example, the use of pretreatment regimes needs to be determined and the effect of such regimes on test material assessed. The standard practice of introducing new antibodies therefore includes running the test material over a range of dilutions with and without pretreatment (heat-mediated, enzyme unmasking, separately or in combination) and assessing the incubation time, either at the laboratory's standard time or, in certain cases, extended to overnight at 4 °C. This extension of the incubation time (with accompanying reduction in concentration, increase in dilution) may be necessary to detect small amounts of antigen.

Similarly, if a new batch of antibody is introduced into the laboratory, where possible the concentration should be checked against the previous antibody batch and its effect on the dilution taken into account when used in a new staining run. It is essential for the working concentration of the biomarker to be known, not the dilution, as the concentration of a biomarker can influence the apparent relationship between biomarker expression and outcome [7], a factor highlighted by Hall & McCluggage [8] as being recognized and largely ignored in the field of p53 ICC for over 20 years.

Using antibodies to Her-2, oestrogen receptor and p53, McCabe *et al.* [7] showed that variation in dilution of the primary antibody can have a profound effect on the clinical significance of an antibody used as a biomarker. In an editorial to this paper, Henson [9] emphasized the variables of ICC, including the concentration of the biomarker, dilution of the antibody, sensitivities of the detection systems, use of proper controls and quality of the reagents.

Warford *et al.* [10] showed that several monoclonal antibodies to JAM-2, CD99, CD138, CD45 and MHC class II antigen gave different expression profiles, dependent upon clone-epitope reaction mechanics. Overall, 19/35 antibodies stained haematopoietic cells by ICC. Moreover, inappropriate cross-reactivities were noted when used on formalin-fixed paraffin sections, compared to acetone-fixed cryostat sections. These authors cited protein conformational differences, cell–cell interaction, post-translational modification, difference in epitope occurrence due to transcribing gene splicing, and differing fixation protocols as playing a role in affecting ICC results.

Any new antibody entering the clinical laboratory therefore should be properly validated prior to being used in the clinical environment. UK NEQAS ICC & ISH recommends that any new antibody should be validated against 50 previously-diagnosed cases. However, depending on validation methods, the testing criteria can vary from 20 to 100 cases [11, 12].

4.2.4 Issues of reproducibility and standardization

Reproducibility between laboratories and between staining batches is essential for good practice. Within one laboratory environment it may be easier to maintain reproducibility, but much of ICC depends on the pre-analytic phase, which involves adequate and complete tissue fixation. Good communication between staff is essential, especially when the ICC laboratory is sited in a different location/building/hospital from that where tissue sampling, fixation and processing are conducted. Good tissue and cell morphology preservation is paramount in achieving good ICC staining, which in turn has a direct effect on the immunophenotyping of tumours.

Simple technical considerations covering the analytic phase of ICC are also essential to standardize reproducibility of staining. Most of these are part of Good Laboratory Practice and includes quality-control procedures on not only the resultant staining pattern, but also different batches of antibodies, buffers, and the use of chemicals and reagents, including the maintenance of pH records for all solutions. These analytic stages of standardizations also encompass the calibration of equipment (automation systems, water bath used in retrieval and so on).

It is therefore important to adopt consistent and reproducible procedures, but at the same time to balance this with the need to be flexible in order to adapt protocols for optimal detection of individual antigens.

Reproducibility between laboratories is more problematic, especially when confronted with different fixation schedules and processing and staining methods. This can be exemplified if we consider different incubation times and dilutions for the same batch of primary antibody, different means of HMER, different buffers (citrate, Tris EDTA, etc.), and post-treatment washes. All these factors contribute to a variation in interlaboratory practice. Add to these differences in supplies between countries and the issue of standardization becomes even more complex. It is therefore advisable to follow the guidelines set out by the commercial suppliers of the antibodies and reagents and to follow the recommendations of bodies such as UK NEQAS ICC & ISH.

Automation systems, along with their standardized and validated protocols, should help in achieving reproducibility and consistency, although variations may still occur between the pre-anlaytic stages of fixation and processing schedules.

Closed automation systems can restrict the use of commercial reagents (antibodies, detection system, buffers, etc.) and are designed to operate within finely set parameters, thus minimizing errors in application but limiting the user in developing new protocols. Closed systems are mainly used by the clinical laboratories, where specific antibodies are standardized using specific controlled protocols. Open systems are mainly used by research laboratories, where there is a need for flexibility to develop protocols.

Automation systems also differ in their delivery of reagent to the tissue section. Reagent delivery can be to a fixed area on the slide, or the slide can be moved to a fixed site of reagent delivery. Systems can be set at more than one incubation temperature or fixed to one temperature. One main advantage of automation over manual staining is the ability to produce staining data, including that required by government regulations for audit purposes. All these parameters come at a higher cost per single test but have the advantages of standardization, reproducibility and decrease in turn-around time for each patient's immunohistochemical test.

Whatever the system being used, some key factors need to be considered in determining systems suitable to a particular laboratory environment. These include reproducibility, the degree of manual attendance required during the staining process, user-friendly software, set-up time and ease of reagent handling.

4.2.5 Control material

Two types of control tissue are generally required to assess that the ICC stain has worked correctly, namely tissue which is known to stain positively for the antigen being used (positive control) and tissue which is known to be negative for a particular antigen (negative control).

4.2.5.1 Positive controls

Positive controls used in ICC contain not only the antigen under investigation but also its normal pattern of distribution. This is important in detecting small amounts of antigen in unknown test material. If the detection system has not been optimized then it is possible for a false-negative result to occur. It is usual to use one positive control section (or cell preparation in cytology) per batch of test slides. Organizations such as the College of American Pathologists (CAP) and Nordic Immunohistochemical Quality Control (NordiQC) advocate the use of mounting control sections on the same slide as each test section. This however can lead to unnecessary duplication and is most appropriate in reference centres, where it would be useful to have known positive material alongside test sections fixed, processed and sectioned in a different laboratory. The supply of positive material is usually conducted in collaboration with clinical colleagues and obtained under the local/national ethical approval system.

New control material systems have been proposed, including the use of peptides attached to the glass slides [13, 14] and antigens attached to beads and embedded in paraffin wax [15]. However, both systems lack the ideal of using tissue architecture to provide the context of antigen distribution and require rigorous investigation as to applicability in the clinical laboratory.

The most commonly used tissue within ICC is either appendix or tonsil, both of which contain cells of different lineage. It is important however to use composite control blocks where appropriate. Composite controls are most often used in breast pathology, where differing levels of antigen expression are included in the block, such as oestrogen receptor (ER) or HER2/NEU.

4.2.5.2 Negative controls

Negative controls take many forms, the use of which depends upon the complexity, applicability and purpose of the test. Essential to any negative control is an understanding of what exactly the negative reaction is telling the user about the method of localization and the tissue under examination.

4.2.5.2.1 Omission of the primary antibody The most common form of negative control is the omission of the primary antibody, replacing it with buffer. Such a negative control should be run with each test section. This type of control tests the detection system for nonspecificity and tests the complete inhibition of endogenous enzyme such as peroxidase. Such a control can also highlight foci of pigment peculiar to the test material that might cause confusion, for example melanin in a suspected melanoma. The major problem with this type of control is the fact that it does not reveal anything regarding the specificity of the primary antibody.

4.2.5.2.2 Immunoglobulins It is important that the subclass of the monoclonal antibody used should be identified. Immunoglobulin-heavy chains are categorized into subclasses, for example mouse IgG is divided into IgG_1, IgG_{2a}, IgG_{2b} and so on. For commercially-available antibodies, this is usually listed on the manufacturer's datasheet. In the case of a purified monoclonal antibody of IgG_1 subclass, a suitable negative control would be the use of mouse immunoglobulin of IgG_1 type, used at the same concentration as the primary antibody. For convenience it is accepted that the unit of antibody concentration be expressed in μg/ml. In the case of polyclonal antisera, immunoglobulin from the same species as the primary antiserum is recommended. Ideally, this control antiserum should be from the same animal prior to immunization with the antigen (pre-immune serum) but this is not usually available to laboratories using commercial sources. 'Normal' serum, i.e. species-specificity, is therefore the best that can be used. These controls show that the test material does not have an affinity for either polyclonal antisera or monoclonal antibodies via complement binding receptors.

4.2.5.2.3 Tissue known to be negative for a particular antigen Another form of negative control uses the primary antibody and complete detection system on material known to be negative for the antigen. Several problems exist with this type of control. It is necessary to have complete certainty that the control tissue does not contain small amounts of antigen below the detection capability of the method. Problems arise when there is a change from one method to another, such as an upgrade in sensitivity in methodologies, when apparently false-positive results may start to appear.

4.2.5.2.4 Blocking controls Competition assays can be used as negative controls. In this case, the immunogen is incubated with the primary antibody. This mixture is then used in place of the primary antibody and the detection system is applied as standard. In theory, the antibody should be absorbed by the immunogen present in high concentrations relative to the primary antibody prior to application to the test material. This may not always result in a complete negative reaction however, as the reaction mechanics may allow for some antibody to preferentially bind with the test material. Most laboratories would accept a reduction in staining intensity as being effective.

For routine diagnostic work, most laboratories use a known positive control with each batch of staining. Some also include a negative control by omitting the primary antibody and replacing it with buffer on the test section. With the majority of antibodies purchased from a commercial source, most laboratories rely on the manufacturer's data sheet for specificity. Antibodies can of course be checked for specificity for special purposes, using techniques such as Western blotting.

4.2.6 Quality assurance (QA)

The purpose of quality assurance is for laboratories to receive feedback from their peers as to how their staining quality stands in relation to the expected 'gold standard' for the particular antigen being tested. ICC is constantly evolving, with the introduction of new antibodies, better detection systems, new instrumentation and so on, such that the expected levels of staining may change. The feedback given from quality-assurance schemes should be used to troubleshoot any problems that a laboratory may have with its methods and protocols.

For diagnostic, prognostic and predictive biomarker use, participation in an external quality-assurance system is essential. Several countries and groups of countries have established their own networks. In most cases the quality assurance takes the form of an assessment, whereby participating laboratories are requested to demonstrate a particular antigen on a selected tissue/cell line/cytological specimen. In the UK it is mandatory for all accredited laboratories to participate in the UK NEQAS ICC & ISH system, although the UK-based UKNEQAS ICC & ISH has participants from over 50 countries. Other non-UK quality-assurance schemes include NordiQC and CAP.

Each QA programme normally consists of a panel of expert assessors made up of health-care scientists and pathologists, who give individual participating laboratories feedback in the form of a report outlining what is and what is not acceptable for a preselected antigen. This may include intensity of staining, poor localization, excessive background, nonspecific staining and so on. The authority exercised by the individual quality-assessment schemes depends on their respective quality-monitoring guidelines, and in the case of UK NEQAS ICC & ISH the organizers have to report consistent underperformance, by any UK laboratory, to the National Quality Assurance Advisory Panel (NQAAP), which may ultimately affect the accreditation of the laboratory.

For research purposes, no equivalent system exists. As outlined above, this lack of agreement has resulted in a diverse reporting system throughout the scientific literature, making it difficult to correlate the significance of results between publications [7, 8].

4.2.7 Immunocytochemistry scoring methods

The simplest scoring system distinguishes between a positive (+) and a negative (−) result. Such a system however does not distinguish between differing levels of positivity. For this, the degree of staining can be graded and given a rating of positivity +, ++, +++. Positivity can also be based on the number of positive cells, which is then usually expressed as a percentage of the total cell population or relevant cell type.

Assessing the staining intensity is more difficult than counting numbers of positive cells and is more subjective. Even when different staining parameters are minimized through automation, differences in specimen collection, fixation, detection systems and section thickness can all influence intensity. Moreover, all assessors need to be clear as to what signifies

Table 4.1 A scoring system which accounts for both the staining intensity and the number of cells staining positive [19]

Score for proportion staining	*Score for staining intensity*
0 = No nuclear staining	0 = No staining
1 = <1% nuclei staining	1 = Weak staining
2 = 1–10% nuclei staining	2 = Moderate staining
3 = 11–33% nuclei staining	3 = Strong staining
4 = 34–66% nuclei staining	
5 = 67–100% nuclei staining	

weak, moderate or strong reaction products. Inter-observer agreement can be achieved within study groups, but between study groups variation is inevitable. Some authors have found that assessing only the number of positive cells to be more reproducible than including intensity criteria [16]. The use of virtual microscopy and image-analysis systems may address these issues.

Several systems for manually assessing both staining intensity and positive cell number have been proposed, such as the H or Quick Score [17], and systems that grade weak, moderate and strong into 1, 2 and 3, respectively, multiplied by the percentage of cells stained (1–100%), with the final result being a score range between 1 and 300 (weak, 1 × 1%, to strong, 3 × 100%). These scoring methods have been further refined, for example, for ER scoring, which not only takes the intensity into account but groups together the percentages of positive cells (Table 4.1) [18, 19].

A system with the essential features of good and poor staining has been proposed for the purposes of auditing ICC [20]. Features have been scored to include staining intensity (0–3), uniformity of staining (0,1), specificity of staining (0,1), absence or presence of background staining (0,1) and counterstaining level (0,1), which cumulatively contribute to the scoring system. This system provides a possible total score of 8 for perfect demonstration of the preparation. Using such an internal quality-scoring system can highlight methodological problems in the clinical laboratory over a given period of time.

Levels of positivity, intensity and numbers, or a combination, do not tell us about the distribution of the positivity however. Distribution of the staining pattern depends on the antibody and its expected protein localization within the tissue, which may be cytoplasmic, nuclear, membrane or a combination of these.

Reproducibility in the reporting of ICC needs to be shown, but the enumeration and quantification of salient features needs to be determined by the application of a biologically-relevant denominator such as how many microscopic fields or nuclei need to be assessed. Hall [21] recommended plotting the 'wandering means', i.e. plotting the mean occurrence of the biomarker against an increasing number of denominator factors (fields, nuclei, etc.). Although he was discussing this in the context of quantifying apoptosis, the same principle can be applied to various tumour biomarkers.

In the field of p53 staining, Hall & McCluggage [8] argue for more robust reporting in the literature on the methods used to assess ICC. Such open reporting, they argue, adds value in targeting this key protein in future therapies, and they call for 'clear criteria for the planning, performance, and reporting of biomarker studies', not only for p53 but for all potential biomarkers.

4.3 Clinical immunocytochemistry

4.3.1 Sample collection in diagnostic cytopathology

It is essential that samples are of adequate quality and contain all salient features. The diagnostic cytopathology laboratory undertaking the analysis should have access to a full range of modern techniques either on-site or at a reference laboratory, which can be used for external staining requests. It is now standard clinical procedure to involve a multidisciplinary team (MDT) approach. These clinical MDTs consist of pathologists, oncologists and/or specialists such as radiologists and haematologists. ICC panels cannot be used in isolation, but when combined with morphology, immunophenotyping can provide a definitive diagnosis. The combination of technologies, from histological appearance on haematoxylin and eosin staining, through clinical history, to ICC and molecular profiling, provides a complete clinical service.

Fine-needle aspirates (FNA, see Chapter 2) are being used more frequently in the diagnosis, staging and follow-up of biopsy sampling of suspicious lumps or enlarged lymph nodes. The collection of cells by this method is often the first line of diagnosis. Moreover, supportive studies are more readily available, including cytomorphology, immunophenotyping and molecular criteria.

Collected by an experienced clinician, the cells are usually stained for morphology using methods such as Giemsa. Cells can be washed in saline and stored in a cytofixative prior to cytospin preparation. Several centres report good results using 95% alcohol or acetone as their fixative. The cells can be prepared using liquid-based cytology (LBC, see Chapter 2). This is more expensive than preparing a monolayer of cells by a cytospin method (cytocentrifugation, see Chapter 2), but has been reported to give superior morphology compared to an ill-prepared cytospin. A good cytospin, prepared as a monolayer and fixed adequately, can however provide an invaluable, cost-effective method for routine immunophenotyping. Excess cells stored in an alcohol-based fixative containing polyethylene glycol (PEG) [22] can be used as control material. Cells can also be prepared using PEG in alcohol as a cell protectant covering the cell preparation. The water-resistant layer, once solidified, is stable for weeks, which allows for the transportation of cell preparations or the storage of control material. The PEG layer is removed using 95% alcohol prior to immunolocalization.

The UK NEQAS ICC & ISH cytology module assesses a participant's ability to demonstrate via ICC various antigens, including those for lymphoma, mesothelioma, adenocarcinoma and melanoma. The module despatches cytospins prepared with a layer of PEG as a protectant to participating laboratories.

In the following sections, technical performance data will be presented from UK NEQAS ICC & ISH. For the majority of schemes at assessment, each participant is sent slides from a common source and asked to stain for the antigen suitable for that module. A panel of four experts scores the slides independently out of 5. Collectively, a score out of 20 is given. Unacceptable staining is graded as 9 or below. Borderline staining is scored between 10 and 12, and acceptable staining between 13 and 20.

4.3.2 Carcinoma or lymphoma?

Immunophenotyping using markers of epithelial (e.g. cytokeratins) and lymphocytic (e.g. leucocyte common antigen LCA/CD45) origin can help in the differential diagnosis between a carcinoma and a lymphoma. Combining ICC with FNA samples can provide further assistance in their differential diagnosis. In samples of bladder washings from patients with bladder cancer, it may also be possible to provide the means for the dynamic monitoring of chemotherapy.

Cytokeratins are intermediate filaments which support the structure of epithelial cells. Their specific localization can help in identifying distant metastases, characterizing squamous carcinomas versus adenocarcinomas and distinguishing between sites of possible primary tumours. Characterized by their Moll numbers [22], cytokeratins decrease in molecular weight as their number increases, thus CK1 (68 kDa) is bigger than CK20 (46 kDa) (Table 4.2).

The CD45 molecule, originally called leucocyte common antigen (LCA), is a family of protein tyrosine phosphatase proteins derived from one gene

Table 4.2 Commonly-used cytokeratin monoclonal antibodies and their CK targets

Monoclonal clone number	*Cytokeratins detected*
MNF116	5, 6, 8, 17 and 19
AE1*	10, 13, 14, 15, 16 and 19
AE3*	1, 2, 3, 4, 5, 6, 7 and 8
CAM5.2	7 and 8
LP34	5 and 6
5D3	8/18
C-50	5/18
OV-TL 12/30	7
C-11 + PCK-26 + CY-90 + KS-1A3 + M20 + A53-B/A2 clone mix	1,4,5,6,8,10,13,18, and 19

*AE1 and AE3 are usually combined to give a 'pan-CK' marker

via tissue-specific alternative splicing. It is found on all haematopoietic cells except for erythrocytes.

4.3.3 Cytology cytokeratin

For the demonstration of cytokeratins in cytology preparations, UK NEQAS ICC & ISH cytospins samples are normally prepared from a pleural aspirate from an adenocarcinoma of the lung. Optimal staining for cytokeratin on such a sample should show cytoplasmic staining of adenocarcinoma cells, with a population of unstained/negative cells, usually reactive polymorphs (Figure 4.2). In keeping with all cytological preparations, nuclear morphology can help to identify cell type. The intensity of the hematoxylin counterstain therefore should be sufficient

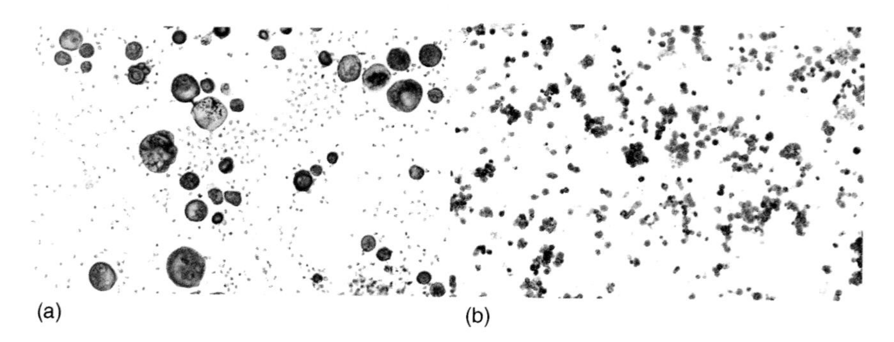

Figure 4.2 Clinical samples of pleural aspirates: (a) adenocarcinoma of lung stained for cytokeratin (MNF116); (b) B-cell lymphoma stained for CD45

Table 4.4 UK NEQAS ICC & ISH general module data CD45 (1.5 years). Although there was no significant difference in the technical performance using these clones to CD45, there was a trend for those participants using X16/99 (CD45), LCA88 (CD45) and UCHL1 (CD45RO) to perform poorly

Clone	*No. of slides assessed*	*Mean score (max. 20)*
PD7/26 + 2B11	2107	14
X16/99	89	12
LCA88	25	12
PD7/26	23	16
UCHL1	6	12
4KB5	5	15
Total	2255	14

the second-most-common type of lymphoma and present as a homogenous group with a t(14:18) genotype and usually a bcl2+, CD10+, bcl6+, MUM1− immunophenotype. Karube *et al.* [29] reported a bcl2−, CD10−, MUM1+ immunophenotype occurring in patients over 55 years of age. The majority of FLs remain indolent for many years but may progress through transformation to diffuse large B-cell lymphoma (DLBCL). Some FLs, however, transform to DLBCL more rapidly. Mutations and promoter-region silencing of tumour-suppressor genes and chromosome abnormalities may advance this progression [30].

The use of FL markers by UK NEQAS ICC & ISH has found that poor immunostaining can result from many factors, including extended protease digestion or antigen retrieval, or conversely insufficient antigen retrieval or digestion, and a too-dilute primary antibody.

4.3.5.1 CD10

CD10 is a transmembrane glycoprotein with metalloprotease activity, which can modulate the effect of bioactive peptides including hormones and growth factors. CD10 is also known as the common acute lymphoblastic leukaemia antigen (CALLA) and is a useful marker of precursor B-cell lymphoblastic lymphoma/leukaemias, follicular lymphomas and Burkitt's lymphomas [31]. It is transitionally expressed on immature lymphocyte precursors of both B- and T-cell type and then re-expressed in proliferating B cells found mainly in the lymphoid germinal centres. It is also detected in mature neutrophil granulocytes and in several epithelial

Table 4.5 General module data CD45 (1.5 years) showing the performance of the participants by clone, based on pass, borderline or fail. It can be seen that users of UCHLI (CD45RO), X16/99 (CD45) and LCA88 (CD45) performed poorly, with 33%, 44% and 52% respectively staining to acceptable levels, whereas 73% of users of the clone mix PD7/26 + 2B11 (CD45RB + CD45) attained acceptable staining. There were too few users of PD7/26 (CD45RB) and 4KB5 (CD45RA) to make any meaningful judgment

	Clone					
	PD7/26 + 2B11 (CD45RB + CD45)	*X16/99 (CD45)*	*LCA88 (CD45)*	*PD7/26 (CD45RB)*	*UCHL1 (CD45RO)*	*4KB5 (CD45 RA)*
Pass (≥13/20)	1538 (73%)	39 (44%)	13 (52%)	21 (91%)	2 (33%)	4 (80%)
Borderline (10–12/20)	405 (19%)	31 (35%)	6 (24%)	2 (9%)	3 (50%)	1 (20%)
Fail (≤9/20)	164 (8%)	19 (21%)	6 (24%)	0 (0%)	1 (17%)	0 (0%)
Total	2107	89	25	23	6	5

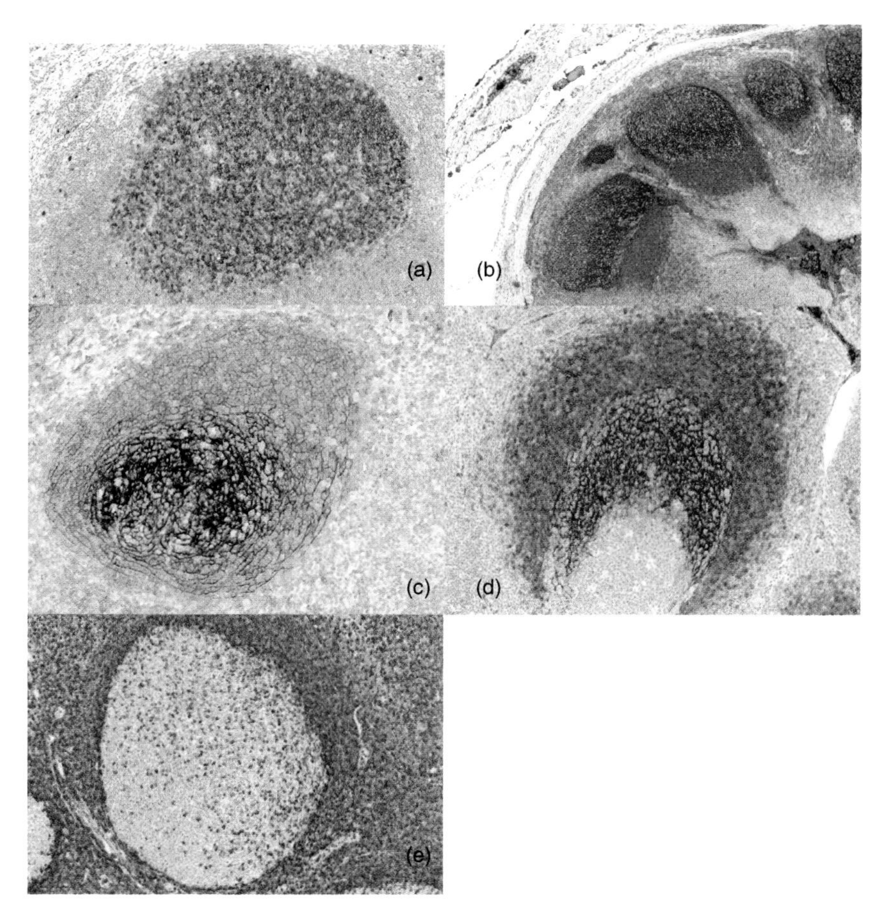

Figure 4.4 Reactive human tonsil sections showing follicular cell staining: (a) CD10; (b) CD21; (c) CD23; (d) IgM. Note the lack of staining in reactive follicular centre cell in bcl2 staining (e)

cell types, including liver, kidney, prostate, intestine and breast myoepithelium, as well as some stromal cells and their tumours [32].

In reactive tonsil, good CD10 staining should show membrane staining in the vast majority of germinal-centre B cells positive with some immature B cells in the peri-follicular region (Figure 4.4a). Conversely, poor staining in a reactive tonsil often shows weak, uneven, missing staining of the relevant cells, or nonspecific staining of cell types or cell compartments not expected to stain. Reviewing the UK NEQAS ICC & ISH data over a six-month period showed that the most common clone was 56C6, which averaged a score of 14/20 (154 slides assessed).

4.3.5.2 CD21

CD21 is a transmembrane glycoprotein and used to be called the C3d complement receptor. In follicular dendritic cells (FDCs) it plays an important role in binding and presenting antigens to mediate affinity selection and maturation of activated B cells [33]. In addition to the FDCs surrounding the germinal centre and the mantle zone of secondary lymphoid follicles, CD21 is also expressed in mature B cells in the marginal and mantle zones (Figure 4.4b). Detection of CD21 by ICC can be a very useful adjunct for detecting structural alterations of the FDC meshwork characteristic in follicular lymphomas, mantle-cell lymphomas, some Hodgkin's lymphomas, angioimmunoblastic T-cell lymphomas, and also for diagnosing the rare FDC sarcomas. Over a six-month period, UK NEQAS ICC & ISH showed that the clones, 1F8 and 2G9, performed equally well, averaging 14/20 and 13/20 respectively.

4.3.5.3 CD23

CD23 is a cell-membrane glycoprotein, also known as the low-affinity IgE receptor, that takes part in the regulation of B-cell maturation. It is primarily detected in B cells excluding early precursors (except if they are Epstein-Barr virus (EBV)-transformed), immunoglobulin-secreting cells, the majority of FDCs, intestinal epithelial cells and some monocytes [34]. CD23 is an important diagnostic marker of B-cell chronic lymphocytic leukemia/small B-cell lymphoma (B-CLL), with the strongest expression in prolymphocytes of the proliferation centres (Figure 4.4c). It can also be useful (along with CD21) for detecting structural alterations of FDCs in some lymphomas and diagnosing FDC sarcomas. Over the same six-month period as above, participants at the UK NEQAS ICC & ISH used three clones to detect CD23, SP23, 1B12 and MHM6.

4.3.5.4 IgM

IgM is a low-affinity pentameric immunoglobulin found in naïve and activated B lymphocytes. FDCs and some B cells in the lymphoid germinal centres, and most mantle-zone B cells of peripheral lymphoid tissues, also contain IgM, as well as some peri-follicular B cells and plasma cells dispersed in these tissues (Figure 4.4d). Over a one-year period, UK NEQAS ICC & ISH data showed that polyclonal antiserum was the most widely used, and participants generally performed better with this than with monoclonal antibodies.

4.3.5.5 Bcl2

The Bcl2 family consists of pro-apoptotic and anti-apoptotic proteins. Bcl2 itself is anti-apoptotic, thus promoting malignant cell survival in response to upregulation as a consequence of the typical t(14:18) translocation found in follicular lymphoma. The most common clone used by participants in the UK NEQAS ICC & ISH scheme is clone 124. In reactive tonsil, good bcl2 staining should show strong cytoplasmic positivity of most lymphocytes except germinal-centre B cells (Figure 4.4e). Some individual cells may show an eccentric ring-like appearance of immunostaining with a wide range of expression. Conversely, poor bcl2 staining is often weak or uneven, or shows nonspecific staining of cell types not expected to stain.

4.3.6 Angiogenesis

Angiogenesis is the process by which new blood vessels are formed. Tumour angiogenesis occurs as the tumour exceeds its original blood supply to avoid hypoxia-induced changes. Angiogenesis plays a vital role in the metastatic process by providing a route by which tumour cells may enter the circulation. Highly-vascularized tumours are associated with a higher incidence of metastasis than less-vascularized tumours [35].

There are several methods of assessing angiogenesis. Non-invasive measurement of angiogenesis has been reported using serum levels of angiogenic factors such as vascular endothelial growth factor (VEGF). In cellular pathology, measurement of angiogenesis is achieved by demonstrating changes in microvessel density (MVD) or microvessel count (MVC). Either MVD or MVC can be assessed using the 'hot spot' method, whereby areas of highest vessel number are counted using a defined objective (X20) or by assessing the leading edge of tumour growth. Reproducibility of assessing blood vessel counts can be improved by experience in identifying hot spots [36] and the use of an eyepiece graticule (e.g. Chalkley methodology), either in isolation or in combination with image-analysis systems. In prostate carcinoma, where MVC matches VEGF staining by ICC, there may be a case for combining anti-angiogenic therapy with conventional treatments [37]. Anti-angiogenic therapies however need to account for the underlying angiogenic pathway and target biology. In such clinical situations, common assessment protocols are required [36]. Moreover, a complete understanding of the basic biology of pro- and anti-angiogenic proteins is essential. For example, alternative splicing during the transcription/translation process leads to protein diversity. The potential for developing markers to various forms of VEGF as a means of understanding the role of pro- and anti-angiogenic forms of VEGF in carcinogenesis [38] cannot be overstated, although whether ICC is the best tool for such studies remains to be determined.

The endothelial-proliferation cell marker CD105 can be used to distinguish between neo-angiogenesis and established blood vessels. Moreover, CD105 immunoreactivity can be a predictor of poor prognosis in non-small-cell lung carcinoma [39], node-negative breast cancer [40] and ovarian carcinoma [41]. Romani *et al.* [42] showed by multivariate logistic regression analysis of 125 patients with colorectal carcinoma that although the number of CD105-positive blood vessels was not correlated with survival, it did identify patients at a higher risk of metastatic disease.

4.3.6.1 Choice of endothelial markers

UK NEQAS ICC & ISH has recorded the use of three endothelial antibodies (CD31, CD34 and VWF) by a total of 394 participants in their assessments for the demonstration of endothelium in appendix control blocks. Optimal staining patterns should show the features outlined in Table 4.6, which were mainly achieved using HMER, protease or trypsin pretreatment.

Table 4.6 Optimal staining patterns to be expected on sections of appendix by angiogenic markers CD31, CD34 and von Willebrand factor VIII (vWF)

Features of good staining, appendix	*Features of suboptimal staining, appendix*
CD31, CD34: strong staining of the endothelial cells in the blood vessels and lymphatic vessels throughout the appendix	CD31, CD34 & vWF: weak or negative staining of the endothelial cells and other elements in the appendix.
CD31: strong staining of the T cells in the base of the lamina propria and weaker staining of the B cells in the follicular mantle with HMER pretreatment	CD31, CD34 & vWF: nonspecific nuclear staining possibly caused by excessive pretreatment
CD34: strong staining on the dendritic interstitial cells in the lamina propria with HMER pretreatment	CD31, CD34 & vWF: inappropriate staining of, for example, epithelium and lymphocytes, probably due to over-pretreatment
vWF: strong staining of the endothelial cells in the blood vessels and lymphatic vessels	
vWF: some background staining of the connective tissue is to be expected	

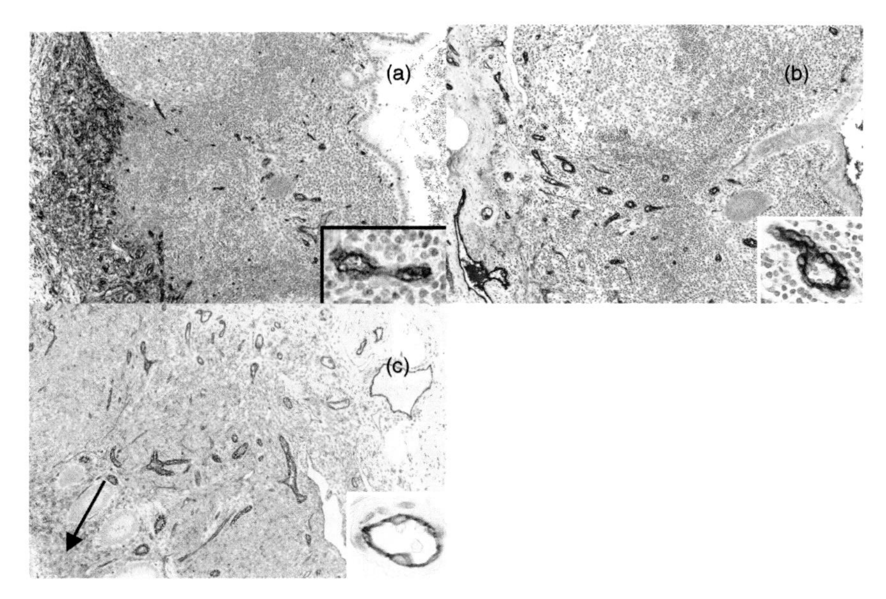

Figure 4.5 Reactive human appendix sections showing blood-vessel staining, insets: (a) CD34; (b) von Willebrand factor; (c) CD31. Note the positive staining of lymphoid cells (arrow) in the CD31 preparation. Insets show blood-vessel staining

Although not statistically analysed, the UK NEQAS ICC & ISH data indicates an overall trend for best participant performance using the CD34 antibody (Figure 4.5a), which has a higher mean score of 15/20, followed by vWF (14/20, Figure 4.5b) and then CD31 (12/20, Figure 4.5c). Participants to the scheme used the JC/70A clone to demonstrate CD31, and the clone QBend10 to demonstrate CD34. Both a monoclonal clone F8/86 and a polyclonal antiserum were used by participants to demonstrate von Willebrand factor VIII.

4.3.7 The cell cycle

Normal cell growth is a tightly regulated process, with the cell going through four phases or stages, each of which is controlled as a checkpoint. Prior to DNA synthesis, the cell proceeds through G1 and the G1/S checkpoint to S phase, where DNA synthesis takes place. The G1/S checkpoint is dependent upon factors such as nuclear size, and environmental growth factors such as nutrient supply. Following DNA synthesis, cell growth continues through G2, the G2/M checkpoint and into M, or the mitotic phase, before re-entering G1 or into a quiescent G0 phase. Kinase proteins associated with specific cyclins form cyclin-dependent kinases

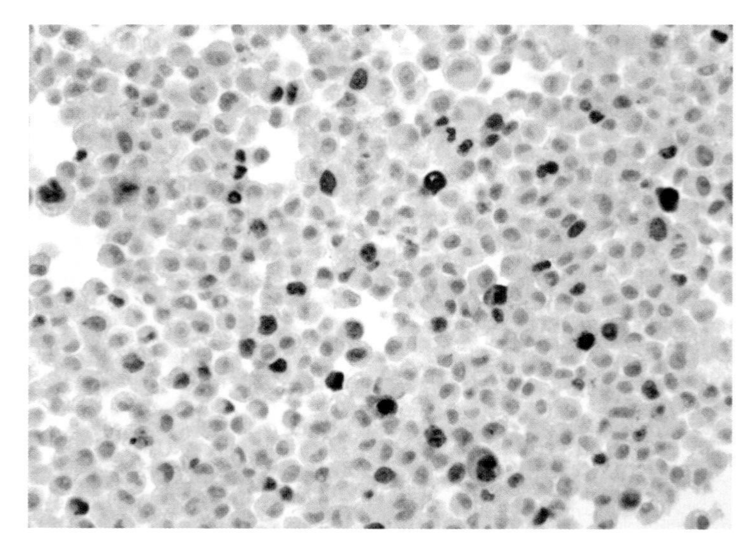

Figure 4.6 Non-small-cell lung-carcinoma cell line stained for the proliferating cell marker Ki67

(CDKs), which control each stage of the cell cycle. These activating cyclins differ in their expression throughout the cell cycle. Either these CDK checkpoints allow the correction of aberrant cell cycle growth or the cycle is aborted, entering apoptosis.

Abnormally high cell proliferation has been identified as having prognostic significance in a number of tumours. Assessing cell proliferation can be done by measuring the number of mitoses or the occurrence of nuclear Ki67 protein (Figure 4.6). Ki67 is a DNA-binding protein which influences chromatin structure [43]. Torp [44] showed that Ki67 indices were associated with a poorer prognosis in astrocytomas.

Results from the UK NEQAS ICC & ISH neuropathology module, which looked at the technical use of Ki67 (Clone MIB1) in a compound block composed of a glioblastoma (high reactivity) and meningioma (low reactivity), showed that optimal nuclear staining relied upon good antigen retrieval. Where this was inadequate or excessive, weak or nonspecific staining resulted, respectively.

4.3.8 Biomarkers predictive of response to therapy

4.3.8.1 Breast: hormone receptor to oestrogen (ER)

Hormonal influence on tumour growth has been known about since the late nineteenth century. With the introduction of the use of an ER inhibitor, tamoxifen in the late twentieth century, and the establishment of methods to assess receptor status, the clinical need for a cost-effective reproducible method was identified. The introduction of the mouse monoclonal clone 1D5 in 1991 for ICC on paraffin-embedded wax sections to replace the Abbott charcoal methodology was quickly validated. Insensitive immunohistochemical assays for ER, however, have been reported to be due to inefficient HMER [45, 46]. More recently, the rabbit monoclonal antibody clone SP1 has been reported to be more sensitive than 1D5 when a dextran-coated charcoal ligand-binding assay is used as the gold standard [47]. However, as noted by Dowsett [48], the advantage of SP1 may not necessarily extend over to the more commonly-used 6F11 clone. Ibrahim *et al.* [49] have found that methodology does indeed matter, and in their UK NEQAS ICC & ISH assessment of the rabbit monoclonal progesterone antibody, SP2, there was a tendency towards false-positive staining by participants. As previously stated, not only should any new antibody introduced into the lab be properly validated but proper control tissue should be used. UK NEQAS ICC & ISH recommends that a good control slide for ER status should comprise a composite block of infiltrating ductal breast carcinoma samples, showing: (i) >80% tumour positivity with high intensity (Allred/Quick score 7–8); (ii) 30–70% tumour positivity with low–moderate intensity (Allred/Quick score 4–6); (iii) negative tumour (Allred/Quick score 0). There should also be normal glands within the tissues that can be used as an internal control.

For breast hormonal receptors the nuclear positivity expected by ICC is scored according to the number of positive tumour nuclei and the intensity of staining. A threshold of the number of positive tumour nuclei is used. This is usually 10% of tumour nuclei, although some centres use more than 20%. This lack of standardization is inhibiting the development of assays and the identification of the prognostic value of assaying for the second receptor ERβ [50]. ERβ has shifted our understanding of ER biology, where its dependence on ERα with co-regulatory proteins and growth pathways means that it is now recommended that both ERα and ERβ are considered together and not in isolation [51].

Results from UK NEQAS ICC & ISH indicate that the 6F11 clone is the most widely used (57%), followed by the 1D5 (27%) and then the SP1 clone (16%) (Table 4.7). Data from the breast hormone module confirms that there is a slight improvement in technical use of the rabbit monoclonal antibody SP1 over the mouse monoclonal antibodies 1D5 and 6F11, although this is not statistically significant. All three clones (6F11, 1D5 and SP1) recognize ERα. The UK NEQAS ICC & ISH data on ER was

Table 4.7 Results from UK NEQAS ICC & ISH breast hormone module ER. A comparison of the performance by monoclonal clone types 2004–2006. All scores ranged from 4 to 20

Clone	*No. of slides assessed*	*Mean score (max. 20)*
6F11	1768	13
1D5	841	11
SP1	516	14
Total	3125	13

collected from 2004–2006 using FFPE sections of breast carcinomas of different levels of ER expression. These were mounted on one slide (Figure 4.7).

4.3.8.2 Breast: ERBB2 (HER2/NEU)

The ERBB2 gene (HER2 in human and NEU in rodents) encodes for a membrane tyrosine kinase receptor. It is a member of the epidermal

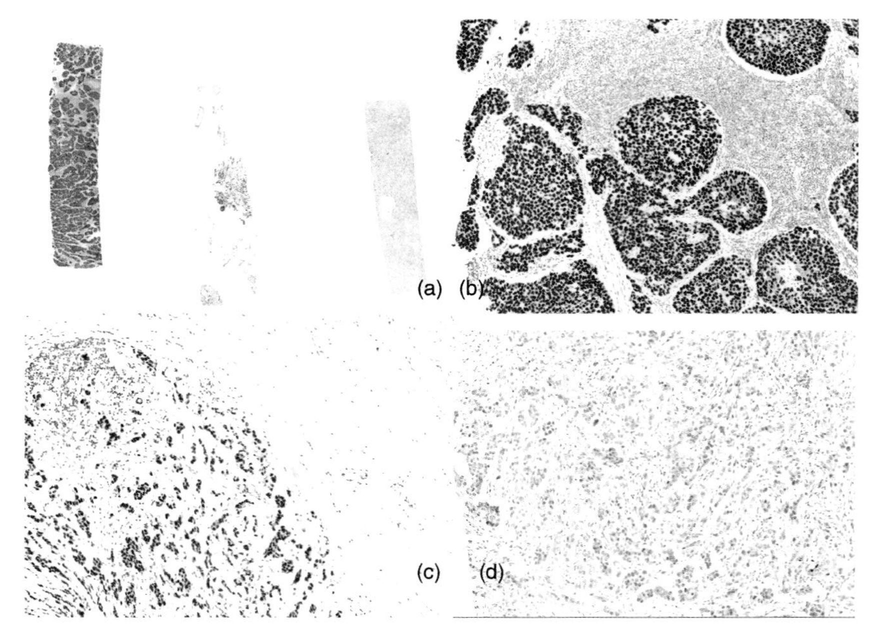

Figure 4.7 Breast carcinoma sections showing (a) composite block of (b) high ER; (c) medium ER; (d) low ER immunoreactivity

growth-factor receptor (EGFR) family, a family of receptor tyrosine kinases which form homo- or hetero-dimers and transmit growth signals from the external cellular environment in response to receptor-ligand activation. This family comprises at least four members and is a target for therapy using kinase inhibitors or monoclonal antibodies (for review, see [52]).

In the UK an annual caseload volume of at least 250 cases is recommended for laboratories providing a clinical testing service. Overexpression of HER2 protein is related to the amplification of the HER2/NEU gene locus localized using ISH techniques.

There are five antibodies that are commonly used in the immunohistochemical detection of HER2 membrane staining, ranging from rabbit polyclonal, mouse monoclonals (CB11 & TAB250) to the newer rabbit monoclonals (SP3 & 4B5). All these antibodies are directed against the internal domain of the HER2 protein, with the exception of the TAB250 antibody, which recognizes the external domain.

The data shown in Figures 4.8 and 4.9 is from UK laboratories participating in the UK NEQAS ICC & ISH HER2 ICC scheme. The data was collected in 2004–2006. Over 70% of participants used a kit, namely the Hercep test kit (Dako, Ely K5204, 5, 6 and 7). A further 17% of participants used the A0485 polyclonal antiserum with their own standardized protocols. Ten percent of participants used the mouse monoclonal CB11 clone.

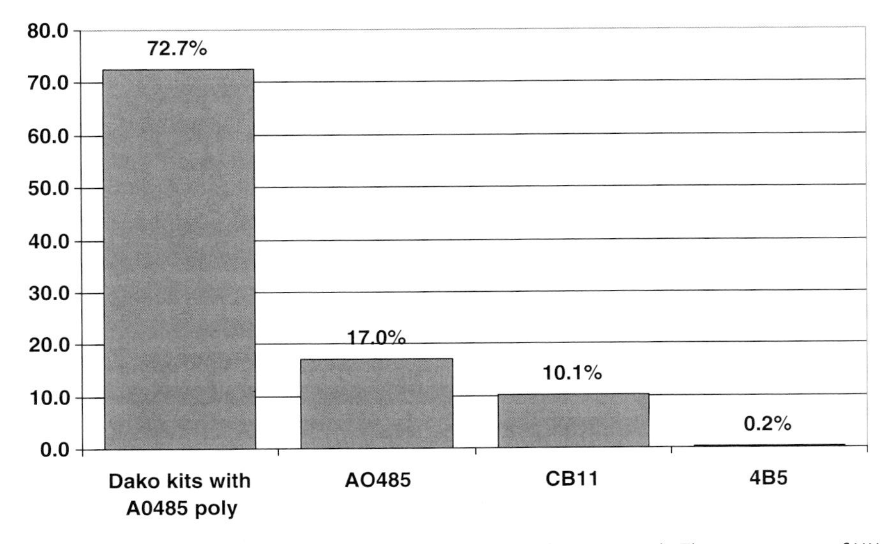

Figure 4.8 Breast module, HER2. UK NEQAS ICC & ISH (2004–2006). The percentage of UK participants using any particular antibody kit, polyclonal antiserum or monoclonal antibody during UKNEQAS ICC & ISH assessments 2004–2006

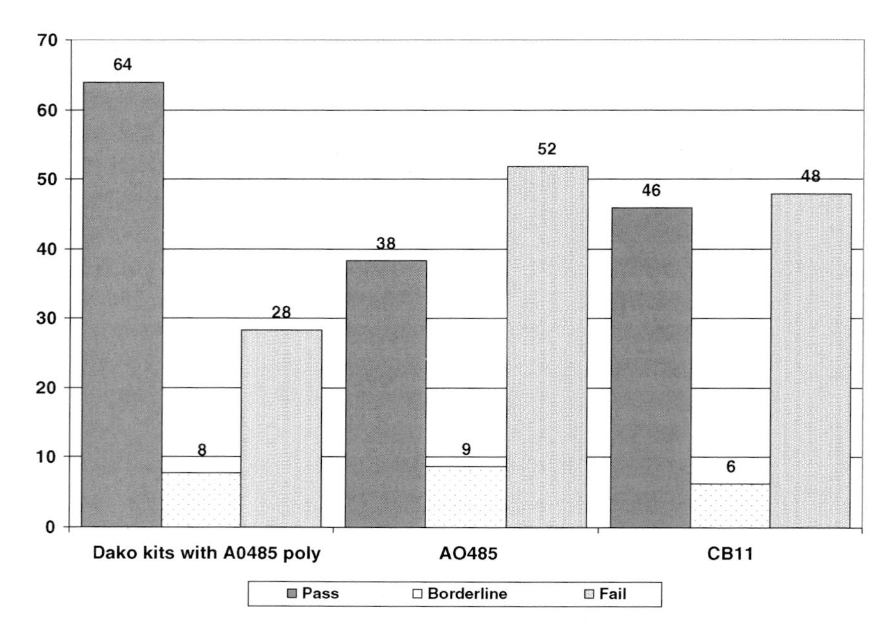

Figure 4.9 Breast module, HER2. UK NEQAS ICC & ISH (2004–2006). Combined pass rates for participants. Note that one participant used a 4B5 clone, which has been excluded from the figure for clarification

At the time of writing, newer rabbit monoclonal antibodies have also started appearing in laboratory repertoires, although they accounted for a very small percentage (0.2%) during this audit period.

The UK NEQAS ICC & ISH HER2 module assesses participating laboratories as having passed (where at least three out of the four assessors agreed that the staining was clinically diagnosable), achieved a borderline pass score (where two of the four assessors agreed that the staining was clinically acceptable) or failed (where three or more of the assessors felt that the staining could not be clinically interpreted). Based on the above guidelines, and ignoring the one participant who used the 4B5 clone, it is quite clear that participants using a standardized kit such as the Hercep test kit performed much better (68% pass rate) than those making up their own in-house protocols, even when using the same A0485 polyclonal antiserum (38% pass rate) that is present in the Hercep test standardized kit.

Unlike in ER staining, where a higher heat-retrieval method, such as pressure cooker, is preferred, the delicate membrane-staining of HER2 ICC is fragile and so a water-bath retrieval method is preferred. Furthermore, the FDA-approved Hercep test kit's method of retrieval cites the water bath,

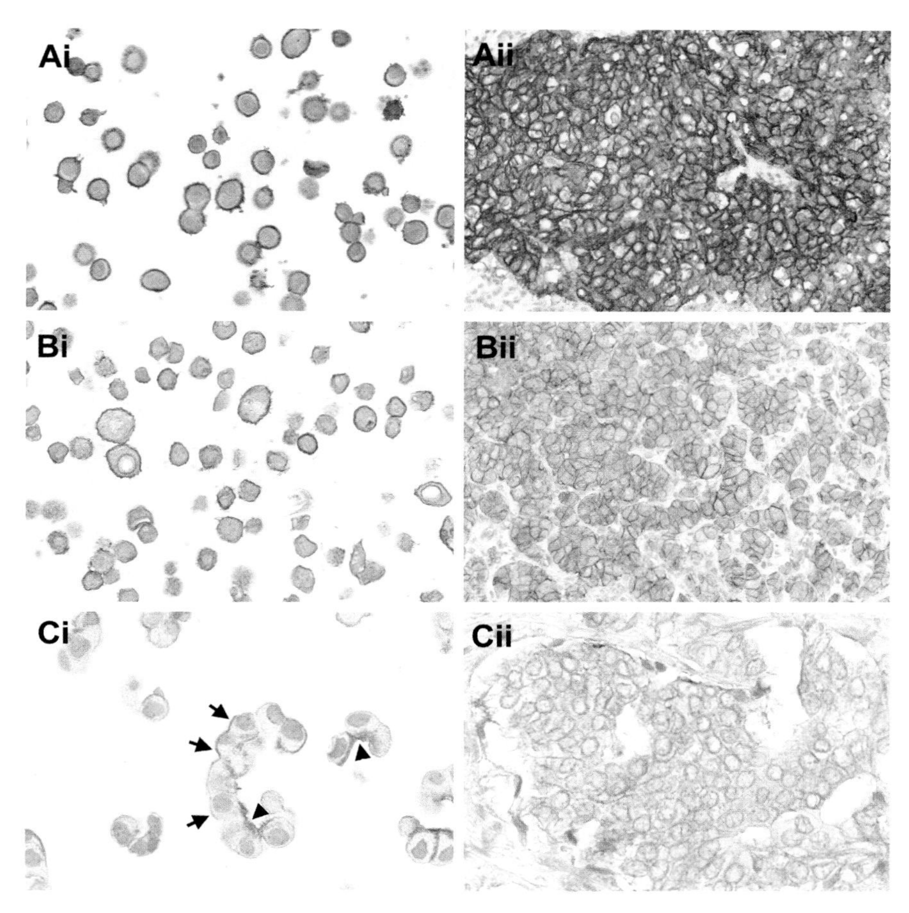

Figure 4.10 HER2 IHC staining of UK NEQAS ICC & ISH HER2 cell lines (Ai, Bi and Ci) and equivalent staining of invasive breast carcinomas: (Ai) 3+ staining on SK-BR-3 cell lines. There is complete and intense staining of the cell membrane, with very little cytoplasmic staining; (Aii) 3+ invasive breast carcinoma showing similar staining to the UK NEQAS ICC & ISH 3+ cell line; (Bi) 2+ staining on MDA-MB-453 cell lines. These cell lines demonstrate complete membrane staining with moderate staining intensity; (Bii) 2+ invasive breast carcinoma, showing similar membrane staining and intensity as the 2+ UK NEQAS ICC & ISH cell line; (Ci) the UK NEQAS ICC & ISH 1+ MDA-MB-175 cell line tends to clump together to form a lumen-like opening, with internally-stained 'brush borders' (arrowheads), which are not scored during assessments. Partial incomplete membrane staining (arrows) is shown on the outer margins of the clumped cells; (Cii) 1+ breast carcinoma showing barely perceptible membrane staining

so this would go hand in hand with the majority of participants using the Hercep test kit. At the time of writing, two new HER2 ICC tests have been launched by Ventana and Leica Microsystems, which use automated systems to carry out standardized onboard retrieval methods.

HER2 ICC is assessed by judging the degree of complete membrane staining of invasive tumour cells and is classed as being: 0 for no staining at all or very slight partial membrane staining in less than 10% of tumour cells; 1 + for faint barely-perceptible membrane staining in more than 10% of tumour cells and/or cells only being stained in part of their membrane; 2 + for weak to moderate complete membrane staining observed in more than 10% of tumour cells; 3 + for strong complete membrane staining in more than 30% of tumour cells (Figure 4.10).

4.3.9 Alimentary tract

Mutation of the c-kit gene and resulting c-kit/CD117 tyrosine kinase receptor immunoreactivity is a feature of soft-tissue tumours arising within the gastrointestinal tract, known as gastrointestinal stromal tumours (GIST). Although originally thought to be smooth muscle tumours, CD34 positivity indicated that they were a separate entity. For reviews see [53] and [54]. The tyrosine kinase inhibitor Glivec (Novartis) is used to treat such tumours, and as such, c-kit/CD117 immunoreactivity has been advocated as a biomarker predictive of response to this form of tyrosine kinase-targeted therapy.

4.3.9.1 Heat retrieval or no heat retrieval?

The consensus meeting for the management of GISTs indicated that approximately 5% of histologically-suspected GISTs are c-kit/CD117 negative and that immunohistochemistry should be performed without antigen retrieval as HMER might result in false-positive staining. Furthermore, a recent publication by Parfitt *et al.* [55] also indicates the potential for misdiagnosis of gastrointestinal Kaposi sarcoma (KS) as GISTs, following HMER prior to CD117 immunocytochemistry.

At the time of writing this chapter, the UK NEQAS ICC & ISH pilot module for c-kit/CD117 requests participants to carry out staining on two separate tissue sections, using HMER on one section and non-HMER on the other. Participants are requested to adjust the antibody dilution so that the distributed GIST sections are adequately stained with both HMER and non-HMER methods.

The UK NEQAS ICC & ISH GIST pilot study data in Table 4.8 shows that participants using HMER performed on average (mean score 14/20) better than those using non-HMER (mean score of 10/20) (Table 4.9).

Table 4.8 Results from the UK NEQAS ICC & ISH alimentary tract module where participants used HMER prior to staining. Data shows a comparison between the main clones used and the mean scores attained. Data from four separate assessments over a two-year period (2004–2006)

Clone	*No. of slides assessed*	*Mean score (max. 20)*
A4502 (rabbit poly)	161	14
T595 (Mouse mono)	13	8
9.7 (rabbit mono)	6	12
Total	180	13

However, it should be noted that the main reason for failure on the non-HMER section was participants not adjusting the antibody dilution. UK NEQAS ICC & ISH advocates the use of optimized primary antibody dilutions for both the HMER- and non-HMER-treated sections. The majority of participants use the Dako A4502 polyclonal antibody.

Ideally a GIST with adjoining intestine should be used as a positive control, or else a GIST with a section of appendix on the same slide. For both HMER and non-HMER sections, there should be good localization of CD117 in the GIST (Figure 4.11). In the adjoining intestine or appendix, CD117 should show good localization of interstitial mast cells and cells of Cajal [56]. Suboptimal staining is normally characterized by weak and/or patchy staining of the GIST, with little or no staining of mast or Cajal cells. Excessive HMER can also lead not only to excessive background but also to nonspecific staining, normally evident in the surrounding epithelium, intestine or appendix.

Table 4.9 Results from the UK NEQAS ICC & ISH Alimentary tract module, stained at the same time as in Table 4.8 but without the use of HMER prior to staining. Data shows a comparison between the main clones used and the mean scores attained. Data from four separate assessments over a two-year period (2004–2006)

Clone	*No. of slides assessed*	*Mean score (max. 20)*
A4502 (rabbit poly)	139	10
T595 (Mouse mono)	10	5
9.7 (rabbit mono)	5	4
Total	154	10

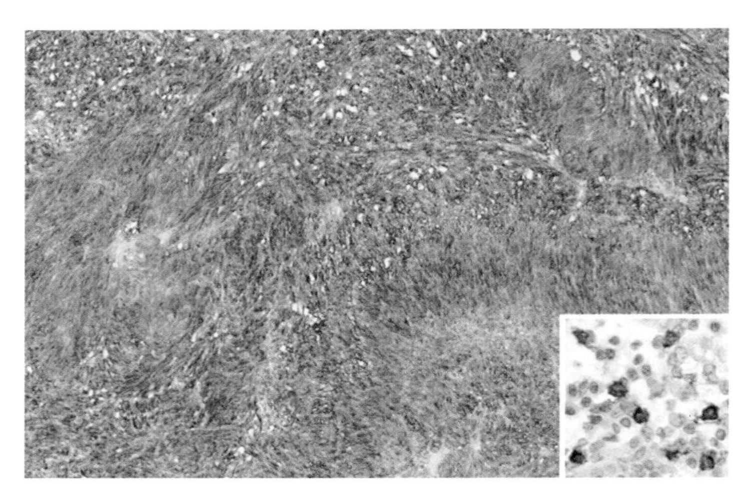

Figure 4.11 Gastrointestinal tumour. CD117-positive following HMER showing strong immunoreactivity and (inset) interstitial mast-cell positivity

4.4 Conclusions

Although ICC is now an established discipline in both the research and diagnostic cellular pathology settings, it remains the case that lack of standardization of protocols and assessment procedures often results in poor and inconsistent results. Establishing protocols for the workflow through the laboratory can increase consistency, and establishing working relationships with external organizations interested in quality assurance increases good performance recognized by external peer review. We have illustrated how the end-product of the ICC process depends on pre-analytical, analytical and post-analytical steps, the results of which can play an essential role in patient diagnosis and treatment. With the ever-expanding number of biomarkers as targets for therapy, the central role ICC plays in cellular pathology and clinical practice is assured and is guaranteed to demand best practice from scientist and medical practitioner alike.

Acknowledgements

We wish to thank all the assessors of UK NEQAS ICC & ISH for contributing to the compilation of the data presented within this chapter. We also wish to thank Dr Tibor Krenics for his assistance with Section 4.3.5 on follicular lymphomas.

References

[1] Shi SR, Key ME, Kalra KL. Antigen retrieval in formalin-fixed, paraffin-embedded tissues: an enhancement method for immunohistochemical staining based on microwave oven heating of tissue sections. J Histochem Cytochem. 1991;39:741–8.

[2] Morgan JM, Navabi H, Schmid KW, Jasani B. Possible role of tissue-bound calcium ions in citrate-mediated high-temperature antigen retrieval. J Pathol. 1994;174:301–7.

[3] Morgan JM, Navabi H, Jasani B. Role of calcium chelation in high-temperature antigen retrieval at different pH values. J Pathol. 1997;182:233–7.

[4] Yamashita S, Okada Y. Mechanisms of heat-induced antigen retrieval: analyses *in vitro* employing SDS PAGE and immunohistochemistry. J Histochem Cytochem. 2005;53:13–21.

[5] Shi SR, Cote RJ, Taylor CR. Antigen retrieval techniques: current perspectives. J Histochem Cytochem. 2001;49:931–7.

[6] Emoto K, Yamashita S, Okada Y. Mechanisms of heat-induced antigen retrieval: does pH or ionic strength of solution play a role for refolding of antigens? J Histochem Cytochem. 2005;53:1311–21.

[7] McCabe A, Dolled-Filhart M, Camp RL, Rimm DL. Automated quantitative analysis (AQUA) of *in situ* protein expression, antibody concentration and prognosis. J Natl Cancer Inst. 2005;97:1808–15.

[8] Hall PA, McCluggage WG. Assessing p53 in clinical contexts: unlearned lessons and new perspectives. J Pathol. 2006:208:1–6.

[9] Henson DE. Back to the drawing board on immunohistochemistry and predictive factors. J Natl Cancer Inst. 2005;97:1796–7.

[10] Warford A, Flack G, Conquer JS, Zola H, McCafferty J. Assessing the potential of immunohistochemistry for systematic gene expression profiling. J Immunol Methods. 2007;318:125–37.

[11] Wolff AC, Hammond EH, Swartz JNet al. American Society of Clinical Oncology/ College of American Pathologists guideline recommendations of human epidermal growth factor receptor 2 testing in breast cancer. J Clin Oncol. 2007;25:118–45.

[12] Walker RA, Bartlett JM, Dowsett M, Ellis IO, Hanby AM, Jasani B, Miller K, Pinder SE. HER2 testing in the UK: further update to recommendations. J Clin Pathol. 2008 July; 61(7):818–24.

[13] Sompuram SR, Kodela V, Ramanathan H, Wescott C, Radcliffe G, Bogen SA. Synthetic peptides identified from phage-displayed combinatorial libraries as immunodiagnostic assay surrogate quality-control targets. Clin Chem. 2002;48:410–20.

[14] Sompuram SR, Kodela C, Zhang K et al. A novel quality control slide for quantitative immunohistochemistry testing. J Histochem Cytochem. 2002;50:1425–34.

[15] Shi SR, Liu C, Perez J, Taylor CR. Protein-embedding technique: a potential approach to standardization of immunohistochemistry for formalin-fixed paraffin-embedded tissue sections. J Histochem Cytochem. 2005;53:1167–70.

[16] Zlobec I, Lugli A, Baker K, Roth S, Minoo P, Hayashi S, Terraccianno L, Jass JR. Role of APAF-1, E-cadherin and peritumoural lymphocyte infiltration in tumour budding in colorectal cancer. J Pathol. 2007;212:260–8.
[17] Detre S, Jotti SG, Dowsett M. A 'quickscore' method for immunohistochemical semiquantitation: validation for oestrogen receptor in breast carcinomas. J Clin Pathol. 1995;48:876–8.
[18] Harvey JM, Clark GM, Osborne K, Allred DG. Estrogen receptor status by immunohistochemistry is superior to the ligand-binding assay for predicting response to adjuvant endocrine therapy in breast cancer. J Clin Oncol. 1999;17:1474–81.
[19] Leake R, Barnes D, Pinder S, Ellis I, Anderson L, Anderson T, Adamson R, Rhodes T, Miller K, Walker R. Immunohiostochemical detection of steroid receptors in breast cancer: a working protocol. J Clin Pathol. 2000;53:634–5.
[20] Maxwell P, McCluggage WG. Audit and internal quality control in immunohistochemistry. J Clin Pathol. 2000;53:929–32.
[21] Hall PA. Assessing apoptosis: a critical survey. Endocrine-related Cancer. 1999;6:3–8.
[22] Maxwell P, Patterson AH, Jamison J, Miller K, Anderson N. The use of alcohol-fixed cytospins protected by 10% polyethylene glycol in immunocytology external quality assurance. J Clin. Pathol. 1999;52:141–4.
[23] Moll R, Franke WW, Schiller DL. The catalog of human cytokeratins: patterns of expression in normal epithelia, tumors and cultured cells. Cell. 1982;31:11–24.
[24] Prasad RR, Narasimhan R, Sankaran V, Veliath AJ. Fine-needle aspiration cytology in the diagnosis of superficial lymphadenopathy: an analysis of 2. 418 cases. Diagn Cytopathol. 1996;15:382–6.
[25] Simsir A, Fetsch P, Stetler-Stevenson M, Abati A. Immunophenotypic analysis of non-Hodgkin's lymphomas in cytologic specimens: a correlative study of immunocytochemical and flow cytometric techniques. Diagn Cytopathol. 1999;20:278–84.
[26] Hehn ST, Grogan TM, Miller TP. Utility of fine-needle aspiration as a diagnostic technique in lymphoma. J Clin Oncol. 2004;22:3046–52.
[27] Gong JZ, Snyder MJ, Lagoo AS, Vollmer RT, Dash RR, Madden JF, Buckley PJ, Jones CK. Diagnostic impact of core-needle biopsy on fine-needle aspiration of non-Hodgkin lymphoma. Diagn Cytopathol. 2004;31:23–30.
[28] Levine PH, Zamuco R, Yee HT. Role of fine-needle aspiration cytology in breast lymphoma. Diagn Cytopathol. 2004;30:332–40.
[29] Karube K, Guo Y, Suzumiya J et al. CD10-MUM + follicular lymphoma lacks bcl2 gene translocation and shows characteristic biologic and clinical features. Blood. 1997;109:3076–9.
[30] Bosga-Bouwer AG, van den Berg A, Haralambieva Eet al. Molecular, cytogenetic and immunophenotypic characterization of follicular lymphoma grade 3B; a separate entity or part of the spectrum of diffuse large B-cell lymphoma or follicular lymphoma? Hum Pathol. 2006;37:528–33.
[31] Jaffe ES, Harris NL, Stein H, Vardiman JW. WHO classification of tumours: pathology and genetics of tumours of haematopoietic and lymphoid tissues. Lyon: IARC Press; 2001.

[32] Chu P, Arber DA Paraffin-section detection of CD10 in 505 non-hematopoietic neoplasms: frequent expression in renal cell carcinoma and endometrial stromal sarcoma. Am J Clin Pathol. 2000;113:374–82.
[33] Bagdi E, Krenacs L, Krenacs T, Miller K, Isaacson PG. Follicular dendritic cells in reactive and neoplastic lymphoid tissues: a re-evaluation of staining patterns of CD21, CD23 and CD35 antibodies in paraffin sections after wet heat-induced epitope retrieval. Appl Immunohistochem Mol Morphol. 2001;9:117–24.
[34] Pallesen G. B2.5: the distribution of CD23 in normal human tissues and in malignant lymphomas. In: McMichael AJ, Beverley PCL, Cobbold S et al., editors. Leukocyte typing III: white cell differentiation antigens: proceedings of the 3rd international workshop and conference; 1986 Sep 21–26. Oxford: Oxford University Press; 1987; p. 383–6.
[35] Zetter BR. Angiogenesis and tumor metastasis. Annu Rev Med. 1998;49:407–24.
[36] Vermeulen PB, Gasparini G, Fox SB et al. Second international consensus on the methodology and criteria of evaluation of angiogenesis quantification in human solid tumours. Eur J Cancer. 2002;38:1564–79.
[37] Stefanou D, Batistatou A, Kamina S et al. Expression of vascular endothelial growth factor (VEGF) and association with microvessel density in benign prostatic hyperplasia and prostate cancer. In Vivo. 2004;18:155–60.
[38] Ladomery MR, Harper SJ, Bates DO. Alternative splicing in angiogenesis: the vascular endothelial growth factor paradigm. Cancer Lett. 2007;249:133–42.
[39] Tanaka F, Otake Y, Yanagihara K et al. Evaluation of angiogenesis in non-small cell lung cancer: comparison between anti-CD34 antibody and CD105 antibody. Clin Cancer Res. 2001;7:3410–15.
[40] Dales JP, Garcia S, Carpentier S et al. Long-term prognostic significance of neoangiogenesis in breast carcinomas: comparison of Tie-2/Tek, CD105 and CD31 immunocytochemical expression. Hum Pathol. 2004;35:176–83.
[41] Taskirin C, Erdem O, Onan A et al. The prognostic value of endoglin (CD105) expression in ovarian carcinoma. Int J Gynecol Cancer. 2006;16:1789–93.
[42] Romani AA, Borghetti AF, del Rio P, Sianesi M, Soliani P. The risk of developing metastatic disease in colorectal cancer is related to CD105-positive vessel count. J Surg Oncol. 2006;93:446–55.
[43] MacCallum DE, Hall PA. The biochemical characterization of the DNA binding activity of Ki67. J Pathol. 200:191:286–98.
[44] Torp SH. Diagnostic and prognostic role of Ki67 immunostaining in astrocytomas using four different antibodies. Clin Neuropathol. 2002;21:252–7.
[45] Rhodes A, Jasani B, Balaton AJ, Barnes DM, Miller KD. Frequency of oestrogen and progesterone receptor positivity by immunohistochemical analysis of 7016 breast carcinomas: correlation with patient age, assay sensitivity, threshold value, and mammographic screening. J Clin Pathol. 2000;53; 688–96.
[46] Rhodes A, Jasani B, Balaton AJ et al. Study of interlaboratory reliability and reproducibility of estrogen and progesterone receptor assays in Europe. Am J Clin Pathol. 2001;115:44–58.
[47] Cheang MCU, Treaba DO, Speers CH. Immunohistochemical detection using the new rabbit monoclonal antibody SP1 of estrogen receptor in breast cancer is superior to mouse monoclonal 1D5 in predicting survival. J Clin Oncol. 226;24:5637–44.

[48] Dowsett M Estrogen receptor: methodology matters. J Clin Oncol. 2006;24 (36): 5626–8.
[49] Ibrahim M, Dodson A, Barnett S, Fish D, Jasani B, Miller K. Potential for false-positive staining with a rabbit monoclonal antibody to progesterone receptor (SP2): findings of the UK National External Quality Assessment Scheme for Immunocytochemistry and FISH highlight the need for correct validation of antibodies on introduction to the laboratory. Am J Clin Pathol. 2008;129:398–409.
[50] Carder PJ, Murphy CE, Dervan Pet al. A multi-centre investigation towards reaching a consensus on the immunohistochemical detection of ERbeta in archival formalin-fixed paraffin embedded human breast tissue. Breast Cancer Res Treatment. 2005;92:287–93.
[51] Speirs V, Walker RA. New perspectives into the biological and clinical relevance of oestrogen receptors in the human breast. J Pathol. 2007;211:499–506.
[52] Meric-Bernstam F, Hung M-C. Advances in targeting human epidermal growth factor receptor-2 signaling for cancer therapy. Clin Cancer Res. 2006;12:6326–30.
[53] Parfitt JR, Streutker CJ, Riddell RH, Driman DK. Gastrointestinal stromal tumors: a contemporary review. Pathol Res Pract. 2006;202:837–47.
[54] Blay J-Y, Bonvalot S, Casali Pet al. Consensus meeting for the management of gastrointestinal stromal tumors: report of the GIST consensus conference of 20–21 March 2004, under the auspices of ESMO. Ann Oncol. 2005;16:566–78.
[55] Parfitt JR, Rodriguez-Justo M, Feakins R, Novelli MR. Gastrointestinal Kaposi's sarcoma: CD117 expression and the potential for misdiagnosis as gastrointestinal stromal tumour. Histopathology. 2008;52:816–23.
[56] Sciot R, Debiec-Rychterb M, Daugaardc S et al. Distribution and prognostic value of histopathologic data and immunohistochemical markers in gastrointestinal stromal tumours (GISTs): an analysis of the EORTC phase III trial of treatment of metastatic GISTs with imatinib mesylate. Eur J Cancer. 2008;44:1855–60.

5

Microarray-based Comparative Genomic Hybridization

David S.P. Tan, Rachael Natrajan and Jorge S. Reis-Filho
Molecular Pathology, The Breakthrough Breast Cancer Research Centre, Institute of Cancer Research

5.1 Introduction

'Targeted therapy' [1, 2] refers to the rational design and development of therapeutic agents with effects that are particular to the specific molecular features of cancer cells. Its goal is to improve on the efficacy of current anti-tumour treatments whilst minimizing adverse effects on normal cells.

Oncogenes and tumour-suppressor genes are critical for both cell proliferation and cell fate determination (differentiation, senescence and apoptosis), with cell type- and context-specific effects: overexpression of a specific oncogene can enhance proliferation in one cell type but induce apoptosis in another [3, 4]. Owing to selective pressure during tumour development [5], it is believed that cancer cells often become 'addicted to' (i.e. physiologically dependent on) the continued activity of specific activated or overexpressed oncogenes (molecular drivers) for maintenance of their malignant phenotype [1–4], with the subsequent clonal selection of cells harbouring the constitutively activated or amplified oncogene within the tumour. The ability to induce proliferative arrest, differentiation and apoptosis by 'switching off' these genes [1, 2, 6] is therefore a highly attractive strategy for molecular targeted therapy in cancer.

Advanced Techniques in Diagnostic Cellular Pathology Edited by Mary Hannon–Fletcher and Perry Maxwell

Where such 'molecular drivers' exist in tumours, such as the BCR-ABL translocation in Philadelphia chromosome-positive chronic myeloid leukaemia (CML) [7] or KIT in soft gastrointestinal stromal tumours [2], targeted therapy in the form of monoclonal antibodies or small molecule inhibitors, like Imatinib, has been shown to induce clinical responses in patients with tumours harbouring these specific genetic alterations [2]. In the context of solid tumours, this principle of 'oncogene addiction' is illustrated by the improved response rates and survival benefits observed in patients with HER2-positive breast cancers, where overexpression of HER2 is driven by HER2 gene amplification in >90% of cases, following anti-*i* antibody (trastuzumab) therapy [8] and therapy with small molecular inhibitors such as lapatinib [9].

The majority of common solid tumours have complex karyotypes comprising multiple genomic abnormalities with extensive cytogenetic variability [10]. The high level of karyotypic complexity has made a systematic characterization of chromosomal patterns in solid tumours difficult. Hence, whilst many potential targets have been postulated, critical molecular drivers remain elusive, making the development of targeted therapy much more challenging. In the past, one of the main limitations in the process of target identification was the lack of techniques that could help identify recurrent molecular genetic changes in tumours with remarkably complex karyotypes. In recent years, this obstacle has partly been overcome by the development of microarray-based comparative genomic hybridization (aCGH) technologies [11, 12] specifically designed to interrogate the tumour genome for the molecular genetic changes driving tumour growth and progression, including a rapid and detailed analysis of global genomic copy-number changes in entire tumour genomes (Table 5.1). Studies using aCGH have led to new molecular classifications of tumours by comparing patterns of genetic changes [13, 14], improved our understanding of tumourigenesis and tumour progression [15–17], and led to the identification of novel putative molecular therapeutic targets [18–20].

Although progress has been made, and despite the significant swell of aCGH methodologies and data in recent years [12], significant challenges still exist before the promise of aCGH (and its derivatives) can be translated into the clinical reality of targeted therapy discovery. Microarray techniques are subject to considerable data variability, due in part to variations in methods of DNA extraction, probe labelling and hybridization, the type of microarray platform used, the number and biological characteristics of samples analysed, the methods used for microarray and statistical analysis, and results validation [21]. Hence, a whole spectrum of methodological issues need to be addressed

Table 5.1 Comparison of aCGH and other techniques for genetic analysis. +, detectable; −, undetectable; aCGH, microarray-based comparative genomic hybridization; CGH, comparative genomic hybridization; CISH, chromogenic *in situ* hybridization; DM, double minutes; DI, distributed insertions; FISH, fluorescent *in situ* hybridization; HSR, homogenous staining regions; LOH, loss of heterozygosity; MIP, molecular inversion probes; SKY, spectral karyotyping; SNP, single nucleotide polymorphisms

						Amplification	
Technique	*Polyploidy*	*Aneuploidy*	*Copy Number Silent LOH**	*Balanced Translocation*	*Unbalanced Translocation*	*HSR/DM*	*DI*
Cytogenetic banding	+	+	−	+	+	+	−
SKY	+	+	−	+	+	+	−
LOH	−	+	+	−	+	+	+
FISH/CISH*****	+	+	−	+	+	+	+
CGH	−	+	−	−	+	+	+
BAC aCGH	−	+	−	−	+	+	+
Oligonucleotide aCGH	−	+	−	−	+	+	+
SNP chips aCGH	+****	+	+	−	+	+	+
MIP arrays aCGH	−	+	+	−	+	+	+
Solexa sequencing	−/+	+	+	+	+	+	+

*Mitotic recombination/gene conversion and endoreduplication.

**FISH and CISH require specific probes mapping to the genomic region of interest, which has to be specifically targeted. CISH has the advantage of allowing a direct comparison between specific genetic abnormalities and morphological features of the cells.

***Deletions and translocations cannot be easily defined by CISH, given the resolution of the probes and the difficulties in performing and analysing multicolour CISH.

****Triploid and tetraploid states can be identified based on the 'B allele' frequency plots.

Table 5.2 Parameters to be considered in the design of microarray-based comparative genomic hybridization studies

Sample-related
Sample type (e.g. FFPE vs. fresh frozen tissue) and number
Tissue and intra-tumour heterogeneity
DNA yield and quality
DNA amplification bias
Links with clinical databases
Tumour specific aberrations vs. copy number polymorphisms
Platform-related
Reliability and reproducibility
Availability
Cost
Quantity of DNA required for assay
Types of aberrations detected, e.g. allelic vs. non-allelic informative platform
Resolution
Analysis methods
Integration of results with those of other high-throughput methods

before the impact of aGCH on translational research can be fully realized (Table 5.2).

5.2 Principles of array CGH

Array CGH is based on the same principles as metaphase CGH, a technique that has been extensively used for the genomic characterization of a number of solid tumours. Both techniques allow the study of DNA copy-number alterations genome-wide, except that the targets for hybridization are mapped clones in the aCGH technique instead of chromosomes as in metaphase CGH. Array CGH, or matrix CGH, (Figure 5.1) was first described in 1997 [22]. Briefly, the technique involves differentially fluorescently labelled test (normal) and reference DNA, which is co-hybridized to spotted probes on the microarray glass slide. These are subsequently scanned to produce an image of differential signal intensities (i.e. dual channel/colour microarrays). Based on the normalized $\log_2$ ratios for each specific clone, a genome-wide (semi)-quantitative analysis of copy-number changes in a given locus is defined. aCGH allows high-resolution mapping of amplicon boundaries and smallest regions of overlap, and improvement in the localization of candidate oncogenes [12] and tumour suppressor genes, and is only limited by the insert size and density of the mapped sequences used.

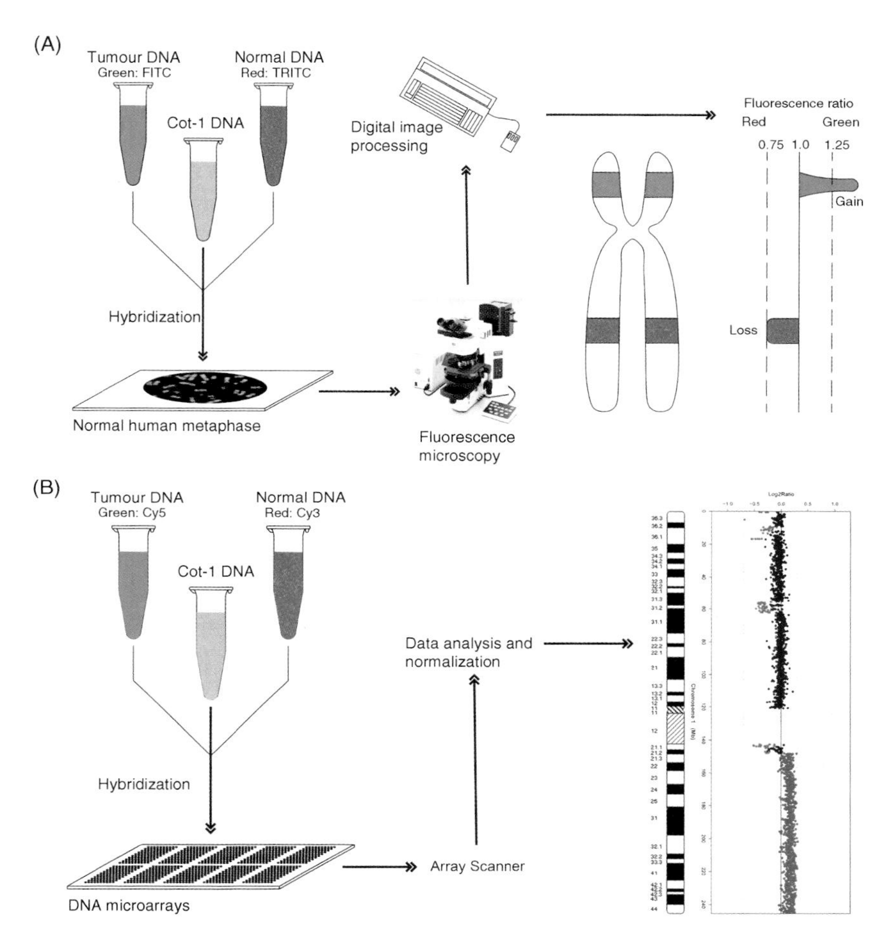

Figure 5.1 Schematic representation of chromosomal and microarray-based comparative genomic hybridization (aCGH). A) CGH: tumour and reference DNA are labelled with green and red fluorophores, respectively, and hybridized to normal lymphocyte metaphase spreads. 15–20 metaphase are captured using a fluorescence microscope coupled with a digital camera. Green to red signal ratios are quantified digitally for each chromosomal region along the chromosomal axis. B) aCGH: as in CGH, tumour and reference DNA are labelled with specific fluorochromes (e.g. Cy3 and Cy5), but then hybridized to glass slides containing bacterial artificial (BACs) with sequence-verified, fluorescent *in situ* hybridization (FISH)-mapped DNA inserts spaced throughout the genome in predefined intervals. After washing, slides are scanned and analysed using specific bioinformatic packages. On the right is shown a diagram of chromosome 1 and a chromosome plot generated with the Breakthrough Breast Cancer 32K microarray platform. Areas of gains and losses are highlighted in green and red, respectively. Note losses on the short arm and a large gain on the long arm (modified from [16])

5.3 aCGH platforms

A wide variety of aCGH platforms are currently available and it should be emphasized that at the moment there is no ideal method for array CGH analysis. Most platforms may be obtained commercially, and they all have advantages and limitations that should be taken into account for study design (Tables 5.1 and 5.2). Two basic types of genomic array technology exist: ordered arrays and random arrays. Ordered arrays are manufactured by spotting (using pins) or synthesizing individual probes in an organized pattern on a planar surface. Random arrays are constructed by immobilizing individual probes on to beads, which are then pooled and assembled on to a patterned planar surface. This allows an average of ~30 replicates of each probe in the array. The identity of each bead is determined following hybridization of specific labelled complements to the probe sequences on the bead. Currently, the majority of aCGH platforms are ordered arrays, with random bead arrays only commercially available from Illumina (San Diego, USA).

Highly-parallel sequencing technologies capable of providing both quantitative (copy number) and qualitative (gene sequence) information from entire tumour genomes in a matter of weeks rather than years have been developed, and are now commercially available. This technology (see Section 5.3.5) has the potential to become the blueprint for the next generation of comparative genomic analysis tools for therapeutic target discovery.

5.3.1 cDNA arrays

Initial studies on genome-wide approaches to aCGH were performed using cDNA microarrays, which were originally designed for expression profiling [23]. The use of this type of microarray for aCGH analysis was due to the availability of cDNA clone sets and the fact that in theory they would allow a direct correlation between genomic deletions and amplifications. However, cDNA microarray analysis enables only the detection of aberrations in known genes and ESTs, since cDNA probes are only representative of expressed genes on a chromosome. The absence of intronic sequences also reduces the stability of the hybridization dynamics, leading to low-hybridization signals, cross-hybridization and reduced sensitivity. In terms of maximal achievable resolution, cDNA arrays cannot compete with currently available alternatives and have already become obsolete in aCGH studies.

5.3.2 BAC arrays

Bacterial artificial chromosomes (BACs), P1-derived artificial chromosomes (PACs) and yeast artificial chromosomes (YACs) are large insert genomic clones that have been widely used in aCGH studies [24–26]. BAC arrays are spotted, dual-channel platforms where test (tumour) and reference DNA are differentially labelled (e.g. Cy3 vs. Cy5) before hybridization on to the array. BAC probes vary in length from 100 to 200 kb and the resolution of each BAC array is defined by the number of unique probes it contains. The probe content of genome-wide BAC arrays ranges from a few hundred to ~32 000 unique elements (tiling path array). Tiling path arrays (i.e. arrays where each BAC overlaps with its contiguous BACs) provide a resolution of up to ~50 kb, given that a genomic change can only be detected if it is sufficiently big to significantly change the hybridization intensity in one of the channels (i.e. change the red:green ratios). These platforms provide sufficiently intense signals for the detection of single-copy-number changes, are able to accurately define the boundaries of genomic aberrations, and, importantly, can be readily applied to DNA extracted from archival formalin-fixed paraffin-embedded (FFPE) tissue as well [27, 28]. One of the main drawbacks with BAC arrays is that their availability from commercial sources is limited and their in-house production is both expensive and highly labour-intensive. In addition, as BAC probes are representative of the human genome, they will also contain repetitive sequences, which can lead to nonspecific hybridization. In order to prevent nonspecific hybridization to these repetitive sequences, Cot-1 DNA is often included in the hybridization reaction, adding to the overall cost of the assay.

5.3.3 Oligonucleotide arrays

Oligonucleotide array aCGH (OaCGH) platforms consist of single-stranded 25–85 mer oligonucleotide elements [26, 29]. Different types of oligonucleotide array have different labelling and hybridization protocols and can provide high-resolution measurements of copy number [26]. There are two main types of oligonucleotide array: single-nucleotide polymorphism (SNP) arrays and non-SNP arrays. Non-SNP arrays include those available from Agilent Technologies (Santa Clara, USA) and Roche NimbleGen (Madison, USA) and comprise 60–75 mer oligonucleotides with site-specific sequences across the genome. SNP arrays, such as those available from Affymetrix (Santa Clara, USA) and Illumina, comprise oligonucleotides that correspond to SNPs along the human genome, and were originally designed for use in linkage analysis and whole-genome genotyping (WGG). Hence, unlike BAC arrays, SNP arrays can also provide information regarding loss of heterozygosity (LOH) and copy-neutral genetic anomalies such as uniparental disomy (UPD) and mitotic recombination, in addition to allelic copy-number changes.

The main limitations with OaCGH platforms are significantly lower signal intensities when compared to those obtained with BAC arrays, and higher probe-to-probe-variation and sequence-dependence of hybridization in the arrays, due in part to greater variation in hybridization dynamics. This is likely to be a function of probe length (oligonucleotide probes are ~100 kb shorter than BAC probes) and GC content, leading to higher variation in signal intensity for similar copy numbers. Lower signal intensities for each probe lead to higher levels of experimental variation/background noise, which render the identification of low-level gains and losses more difficult and the use of degraded DNA (such as that extracted from FFPE samples) challenging. Consequently, although theoretically affording a resolution as high as ~2 kb, signals from several probes (~3–10) need to be averaged or smoothed using specific algorithms before a call can be made. However, the resolution of these platforms can easily be improved by increasing the feature density (i.e. number of SNPs) in the array. Hence, an array with 1000 K SNPs per slide, allowing an averaging of 5–10 SNPs for each call, will allow a resolution of ~10–25 kb.

5.3.4 Molecular inversion probe arrays

Molecular inversion probes (MIPs) are single oligonucleotides with two flanking inverted recognition sequences that recognize and hybridize to specific genomic DNA sequences ranging between 41 and 61 bp in length (Affymetrix). Following probe hybridization to target DNA, a single base-pair gap exists in the middle of the two recognition sequences. This gap can either be an SNP or a nonpolymorphic nucleotide. The gap is enzymatically filled with an appropriate oligonucleotide, resulting in the formation of a circularized probe that subsequently undergoes ‘probe inversion’, which enables the probe to be amplified. Cross-reacted or unreacted probes are separated from the resulting circularized probe via an exonuclease reaction. Each MIP oligonucleotide has a unique sequence barcode tag, which can be assayed via a tag microarray once it anneals to its specific complementary genomic sequence and is circularized [29].

MIP technology has several theoretical advantages over other array platforms, including: higher probe specificity and performance, and thus the robustness of genotype and copy-number determination; a reduced risk of amplification bias and overall cost of the assay, since no PCR amplification is required at the point of mutation detection; and greater flexibility in terms of designing MIP probes, as signals are assayed using a tag array, hence affording one the choice of using any unique sequence, specific exons or other interesting sequences in the array. MIP arrays are therefore particularly well-suited to identifying genomic deletions at a very high resolution, for example at exonic and microsatellite marker-level changes [30], and have been reported to work with DNA from FFPE tissue as well [30, 31]. However, MIP assay technology has not yet been extensively tested, in particular because it initially required 2 ug test DNA [31]. A new protocol [32] that requires 75 ng of total genomic DNA and allows for higher precision, lower false-positive rate and accurate absolute copy-number determination of up to 60 copies has been developed very recently, making this technology more attractive for translational research studies.

5.3.5 Solexa sequencing technology

The human genome reference sequence cost an estimated £500 million to produce and an individual human genome analysed today using the Sanger method would cost about £5million [33]. Further cost reductions are essential before the concept of personalized genome sequencing can be realized. Highly-parallel sequencing technologies that can provide both quantitative and qualitative assays of the human genome sequence at a fraction of the cost are being developed [29, 33–38]. Several systems are currently available, including the Illumina 'Solexa' Genome Analyzer, the ABI SOLiD (Sequencing by Oligo Ligation and Detection) next-generation genetic analysis system, the Roche (454) GS FLX sequencer and the Helicos true single molecule sequencing. Each of these methods has advantages and disadvantages related to the length of the sequences generated and the flexibility of the platform. This field is evolving at unprecedented pace and the reader is referred to two recent reviews on this topic [34, 38].

For the purpose of this chapter, we will briefly discuss the Illumina Genome Analyzer (Solexa), which uses four proprietary fluorescently-labelled modified nucleotides to sequence the millions of clusters of genomic DNA present on a flow-cell surface. Following fragmentation, test DNA is ligated to adapters that facilitate the binding of the single-stranded DNA fragments to the flow-cell surface. These immobilized DNA fragments serve as sequence templates. Following this, unlabelled nucleotides and enzyme are added in order to initiate solid-phase bridge amplification. This step creates up to 1000 identical copies of each template in close proximity (diameter of one micron or less). The four labelled nucleotides (adenine, thymine, guanine and cytosine), which possess a reversible termination property, are then introduced. This allows each cycle of the sequencing reaction to occur simultaneously in the presence of all four nucleotides. At the end of each cycle, laser excitation results in emitted fluorescence from each cluster on the flow cell, following which the image is captured. This allows the sequence of bases in a given DNA fragment to be acquired a single base at time. Following repeated sequencing cycles, the sequencing data is aligned and compared to reference DNA in order to identify sequence differences as well as copy-number changes for each DNA fragment, using a proprietary genome analyser software package.

The potential to elucidate the global genomic copy-number changes, together with all the activating and inactivating mutations in known oncogenes and tumour-suppressor genes, and the presence of translocations and gene fusions in an entire human tumour genome, is a tantalizing prospect [34, 38]. From a molecular genetics perspective, this technology has the potential of solving several longstanding controversies, such as the mechanism leading to amplification, and rearrangements involved in the genesis of amplified regions. Furthermore, next-generation sequencing allows for the identification of structural genomic aberrations. In fact, we anticipate that this technology will lead to the identification of numerous oncogenic fusion genes in common types of carcinoma [39], contrary to the current belief that fusion genes, although highly prevalent in leukaemias/lymphomas, paediatric tumours and prostate cancer, would be rare in breast, lung and colon cancer. The development of analytical tools and bioinformatics methods to support this type of technology will be undeniably challenging, but also of paramount importance. When available, these technologies may be the very epitome of the ideal profiling tool for the prospect of genomically-tailored therapy.

5.4 Choosing the right platform

The choice of platform is dependent on the types of sample available as this has a direct impact on the quantity, quality and purity (i.e. the proportion of DNA belonging to the cells of interest) of extractable DNA for analysis. The vast majority of translational research studies successfully incorporating aCGH analyses have used DNA extracted from fresh-frozen tissues to provide the highest-quality nucleic acid for analysis. However, the most widely available resource for DNA remains 'locked' in archival FFPE material, which is accompanied by a wealth of clinical follow-up data. Extracted DNA from FFPE is often heavily cross-linked, degraded, fragmented and heterogeneous (i.e. a mix of cells of different genomic composition), and is therefore suboptimal for microarray analysis [40]. Consequently, aCGH profiles of FFPE material generally have larger variances, lower signal intensities and lower dynamic range than hybridizations of fresh-frozen tissue and cell line-derived DNA.

Currently, the majority of aCGH studies that have reported success in using DNA extracted from archival FFPE cancers to identify copy-number changes and putative therapeutic targets have been based on BAC array platforms [29, 20, 41–43]. Nonetheless, there have recently been limited reports of success in aCGH profiling of FFPE tumours using cDNA arrays [44, 45] and Affymetrix [46] array platforms as well. Furthermore, a multiplex-PCR-based quality control procedure that can predict the viability of the test DNA for the aCGH analysis has been described [40]. When using FFPE or fresh-frozen tissue for aCGH analysis, careful microdissection of tumour from surrounding stromal-component tissue to ensure a 70–75% purity of tumour DNA content has been shown to be sufficient [12, 40, 47]. Although bioinformatic algorithms to deal with samples containing >50% non-neoplastic cells have been described [48], the accuracy of the profiles obtained with these approaches remains to be determined.

With laser-capture microdissection, it is now possible to study the genetic features of limited numbers of cells or small lesions of interest. The limiting step for coupling laser-capture microdissection or other microdissection techniques with aCGH has been the small amount of DNA retrieved using these methods. In the study of tumours, where most diagnoses are currently made on core needle biopsies, for example breast cancer, the lack of material can present a significant obstacle to detailed molecular analysis. Although this problem may be overcome by increasing the yield of DNA for aCGH with whole-genome amplification (WGA) methods [28, 49–55], most WGA methods tested to date introduce significant genomic biases in the analysis. However, this can be minimized by ensuring similarly-amplified reference and test DNA samples are used in the assay [49, 56].

Evidently, a platform with higher resolution is likely to provide a more comprehensive picture of the global genomic aberrations in a tumour. For example, the use of high-resolution BAC array analysis has revealed a greater deal of complexity than was previously anticipated. When the 8p12-p11.2 amplicon [57], found in 10–15% of breast cancers and arguably associated with the outcome of breast cancer patients [58], was subjected to high-resolution aCGH analysis, its complexity emerged. Currently it is accepted that this amplicon is composed of at least four distinct amplification cores [59]. Similar findings have been described for other amplicons in breast cancer, including the recurrent amplification on 20q13 [60].

Clearly, the ease of detecting any particular genomic copy-number aberration is inversely proportional to its size (length) and the number of elements involved. Hence, a large 1 Mb region with multiple copy-number gains would encompass multiple elements and be detected by most array platforms, whilst a small 20 kb region with only a single copy-number change (e.g. microdeletion) would be beyond the resolution of most array platforms, with the possible exception of high-density oligonucleotide arrays and SNP arrays, MIP arrays and high-throughput Solexa sequencing. Obviously, cost is another major consideration, and if high-resolution arrays are unavailable one might even choose to combine a lower-resolution BAC array for global genomic analysis with fine-mapping of an individual chromosome or genomic region of interest using custom designed oligonucleotide arrays (e.g. NimbleGen) [61].

Regardless of resolution, however, some platforms are better at picking up specific anomalies than others. Hence, if a global picture of large-scale gains/amplifications and losses is all that is required, any array platform will suffice, provided it is of the desired resolution. Alternatively, if the detection of more subtle aberrations involving copy-number-neutral allelic imbalances such as UPD and mitotic recombination is required, SNP arrays and the Solexa sequencing flow-cell technology are the platforms of choice.

5.5 Analysis and validation

Analysis of micorarray data always poses the statistical problem of false discovery given that thousands of variables (probes/genomic regions) are being investigated using a relatively small number of biological replicates (sample size) due to cost and availability of material [62, 63]. Furthermore, the data cannot be assumed to follow a normal distribution and often many outliers exist. In addition, adjustments need to be made for channel-dependent background and global intensity differences, and also to scale the data.

Various analytical tools and methods [64, 65] have been designed to resolve these problems, and numerous bioinformatics software packages designed for the analysis of aCGH data are publicly available from the World Wide Web. These include regularly-updated versions of the R data-transformation and statistical-analysis program (the R Project for Statistical Computing) and Bio-Conductor (Open Source Software) [66] and some packages better suited for copy-number analysis of SNP chips, such as Genomic Identification of Significant Targets in Cancer (GISTIC) [67]. In addition, all commercially-available aCGH platforms come with their own specific data-analysis software (e.g. Illumina Beadstudio and

QuantiSNP [68]), including platform-specific protocols for data normalization and bioinformatics support. However, a preliminary analysis carried out by our group, comparing Beadstudio and other analysis platforms, revealed the rather low sensitivity of the former [69]. Essentially, these analytical software packages use algorithms to reduce the experimental variation for regions with similar copy numbers (i.e. smoothing algorithms), such as adaptive weighted smoothing (aws) [70], maximum-likelihood models, hidden Markov models [71], Gaussian smoothing [72] or circular binary segmentation [73], before confidently defining genomic changes. In simplistic terms, these analytical methods transform aCGH data by organizing a user-defined consecutive sequence of adjacent signals into regions of constant copy number known as segments, which are subsequently classified as 'gain', 'amplification', 'loss' or 'no change' depending on their signal intensity and tumour:reference signal ratio.

Whichever method is utilized in the analysis of aCGH data, it is vitally important that the invariably and often excessively large volume of data generated is appropriately curated and validated with *in situ* or other molecular methods. These include quantitative real-time PCR (QPCR) techniques [74] and fluorescent (FISH), chromogenic (CISH) or silver (SISH) *in situ* hybridization (see Chapter 6) [19, 75–79]. Additionally, in studies where matched normal DNA samples are not available, regions of genomic interest (i.e. recurrently amplified or deleted) need to be cross-referenced with available germline DNA copy-number polymorphism (CNV) databases [80] and validated, preferentially, with *in situ* methods.

5.6 Finding the target

Fundamental to any good experimental study is good design. With the extensive dimensionality of data generated by aCGH studies, good design has never been more crucial in facilitating data analysis and appropriate interpretation of results [12]. When designing microarray studies, there are three commonly adopted approaches, namely class-comparison, class-discovery and class-prediction studies [21]. In addition, aCGH is also often employed in descriptive molecular genetic characterization of tumours and accurate identification of the boundaries of specific amplicons and deleted regions [12, 18, 19, 77]. Although the majority of microarray-based gene expression and aCGH studies in breast cancer have employed one of the above approaches [50, 81–85], the identification of potentially 'druggable', recurrent molecular genetic changes using aCGH is the focus of this chapter.

The challenge of using aCGH as a screening tool for the identification of therapeutic targets is not trivial. Our group and others have approached this challenge in two distinct ways: through top-down and bottom-up studies. The top-down approach involves profiling a group of histopathologically-homogenous tumours using aCGH to identify recurrent genetic changes that might harbour potential oncogenes or tumour-suppressor genes. Here, the emphasis is on discovering individual genetic aberrations and the genes mapping to these regions that have a significant impact on tumour biology. For example, in ER-positive cancers, recurrent amplifications of 8p11.2-8p12 are observed in approximately 10% of cases. Array CGH can be used to define the boundaries of this amplicon and the genes that map to the smallest region of amplification (SRA, i.e. the genomic region that is amplified in all cases harbouring the amplification). It should be noted that with the increasing resolution of aCGH, it has become apparent that most amplicons are much more complex than was previously anticipated and that these amplicons may comprise more than one SRA and, almost certainly, more than one amplicon driver. Both concepts have recently been demonstrated in breast cancers with amplification of 8p11.2-p12 [59] and of 20q13 [60]. In 8p11.2-p12, at least four distinct amplicon cores with different drivers have been described, whereas in 20q13 two distinct patterns amplification have been found, which are not only associated with differing gene expression profiles but also have distinct prognostic significance.

Apart from detailed and diligent aCGH analysis, this process requires overlaying of array CGH and expression array data (or another type of relatively high-throughput expression profiling), protein profiling or RNA interference (RNAi) analysis. If the expression of the genes mapping to the SRA is determined in the same tumours by means of expression arrays, quantitative real-time PCR (qRT-PCR) or immunohistochemistry (if antibodies for all gene products are available), the genes whose expression correlate with amplification can be identified. However, one should bear in mind that for some genes, expression is more pervasive than gene amplification and correlation coefficients may not be as high as expected. Therefore, ruling out a gene as a possible amplicon driver based on a low correlation coefficient may risk excluding potentially biologically-interesting genes or even the actual amplicon driver. One study ruled out many genes as possible amplicon drivers of 8p11.2-8p12, including FGFR1, based on that approach [86]. We advocate a more cautionary approach, ruling out only those genes that are not expressed when their locus is amplified. Given the complexity of the amplicons, this should be based on a diligent overlaying of genomic and expression data, including the complexity of the amplicon cores [59].

After identifying a specific amplicon and its possible/likely driver(s), the next step is to identify a model to test whether cancer cells of similar phenotype depend on any of those genes for their survival (i.e. if the genes to be studied would elicit an 'oncogene addiction' phenomenon in a specific model that resembles the cancer initially studied). The identification of the model can be achieved either by a detailed profiling of cancer cell lines or by interrogating publicly-available datasets of genomic and expression data of specific cell lines similar to the type of tumour that was studied (e.g. [87] for breast cancer cell lines). It is of utmost importance to identify models that not only harbour the amplicon of interest, but also have a phenotype that is similar to that of the cancer studied, as the functions and biological importance of a significant number of oncogenes appear to be context- and cell-type-dependent. Unlike previous beliefs that cancer cell lines would have so many *in vitro* artefacts that could not be used as models for human cancer, recent studies have provided strong circumstantial evidence to suggest that not only are the phenotypes of tumours recapitulated in cell lines, but many of the genomic aberrations identified by aCGH analysis of human tumours are also found in the genomic profiles of cell lines [87–89]. For instance, up to 10% of luminal cancers harbour 8p11.2-p12 and 11q13.3 co-amplifications [19, 58], and two breast cancer cell lines of luminal phenotype also harbour the same co-amplification (i.e. MDA-MB-134 and SUM44 [19, 87]; a subgroup of basal-like breast cancers harbour amplification of the EGFR gene [17, 19] and two basal-like breast cancer cell lines also harbour this amplification (i.e. MDA-MB-468 and BT-20) [19, 87], and the amplicons are remarkably similar (Reis-Filho JS, unpublished results). Once the model has been identified, the biological significance of each gene can be tested by knocking it down by using RNAi methods and chemical inhibition, if the protein product of the gene can be targeted with compounds already available. This approach has led to the identification of putative therapeutic targets in various tumours, such as E2F3 in bladder cancer [90], RAB25 in breast and ovarian cancer [18], IGF1R in Wilm's tumours [77] and FGFR1 in breast cancer [19] (Figure 5.2).

An alternative approach to identifying potential therapeutic targets in a given tumour type, which has recently been adopted by our group, is based on the identification of the genes pertaining to the SRA of recurrent amplicons and subsequent functional profiling of all genes belonging to these regions by high-throughput small-interfering (siRNA) or short-hairpin (shRNA) screens [91]. This approach circumvents the problem of ruling out genes whose expression does not necessarily correlate with gene amplification (as explained above).

The bottom-up approach involves identifying possible therapeutic targets in cell lines of a particular cancer type and then identifying a

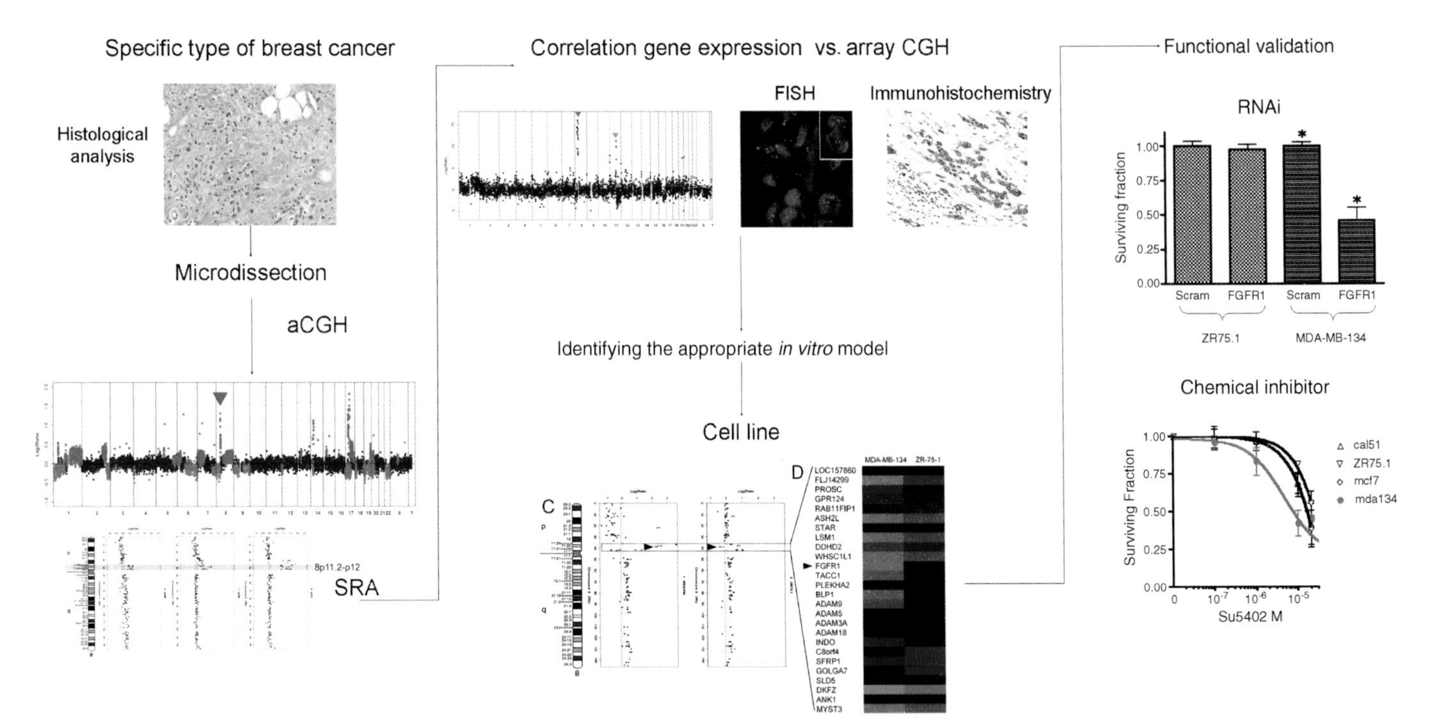

Figure 5.2 Schematic diagram of target identification and validation with array CGH as a discovery tool. After identifying a homogeneous group of cancers based on well-defined criteria, tumours are microdissected and subjected to aCGH analysis. Following profiling and identification of recurrent amplicons (green arrowhead), the smallest region of amplification (SRA) should be defined for each amplicon. This should then be followed by a correlation between amplification and gene (e.g. qRT-PCR or microarrays) or protein expression (e.g. Western blot or immunohistochemistry). After identification of the candidate gene(s), an appropriate *in vitro* model should be identified, based on the phenotypic and genotypic characteristics of the specific cancers studied and the cell lines available. These cell lines can then be used to test the functional significance of the genes of interest (see text). The diagrams demonstrate reduced survival when the oncogene FGFR1 is inhibited with siRNA and a chemical inhibitor in cells harbouring FGFR1 gene amplification and overexpression (modified from [19])

group of tumours with similar phenotypic characteristics that harbour the same genomic changes. For example, by subjecting a set of cell lines to both aCGH and gene-expression array analysis, putative oncogene candidates within a validated region of recurrent amplification can be interrogated at the level of gene expression to identify a shortlist of genes where a good correlation exists between amplification and mRNA overexpression [12, 87]. From this shortlist, Western blots are carried out on cell line lysates to confirm the presence of protein overexpression and its correlation with mRNA expression, whilst parallel immunohistochemical analysis in a larger series of FFPE tissues can be performed to confirm the presence of overexpression of the protein(s) of interest in tumours. The prognostic and predictive significance of these proteins and their association with histopathological subtype will also provide a further layer of evidence with regards to phenotypic relevance. The goal of the analysis of the cell lines is to generate a shortlist of genes that are amplified and overexpressed and which can be tested in clinical samples to determine whether they are of prognostic significance or putative therapeutic targets or both. For example, the analysis of HER2-amplified breast cancer cell lines has demonstrated that other genes pertaining to the HER2 amplicon, such as GRB7 and STARD3, are consistently co-amplified and co-overexpressed with HER2. When these genes are knocked down with siRNA, significantly decreased cell proliferation and cell-cycle progression are observed [92]. These findings provide circumstantial evidence that HER2-amplified cell lines may also be 'addicted' to the signalling of genes co-amplified together with HER2. The recent identification of PVT1 as a gene consistently co-amplified and overexpressed with MYC provides yet another example of this phenomenon, as siRNA reduction in either PVT1 or MYC expression was shown to inhibit proliferation in breast cancer cell lines where both genes were amplified, but not in those lacking amplification of these genes [93]. Tumour DNA can be subjected to gene sequencing to look for mutations of functionally-significant genes as well.

Clearly the type and number of samples available will affect the feasibility of any of the aforementioned approaches, in particular whether a top-down or a bottom-up approach would be more suitable. Where fresh-frozen tumour samples are unavailable and FFPE DNA is not of sufficient quality for aCGH analysis, a bottom-up approach is more likely to be fruitful. Conversely, even with fresh-frozen tissue available, comprehensive molecular genetic profiling of cell lines derived from the tumour subtype of interest has the added advantage of providing investigators with the optimal models for *in vitro* functional assays [87].

5.7 Data integration

One of the challenges of the approaches outlined above is how to define a shortlist of the most likely therapeutic targets. In this respect, integrative analysis of high-throughput data [4–97] from aCGH, gene expression and methylation arrays and proteomic analysis may be particularly useful for several reasons: 1) the impact of methodological unreliability is reduced by cross-validation between data from different biological (i.e. genomic, gene transcription or protein expression) levels; 2) the integration of genomic data (i.e. copy-number changes) with data subject to dynamic changes (e.g. mRNA and protein expression); 3) the facilitation of the development of a systems-biology approach to delineate the interplay between key signalling pathways, networks and regulatory feedback loops that determine disease phenotype. However, such integrative approaches are encumbered by complex logistic and analytical challenges, which include: 1) the immense biological complexity in progressing from genotype to phenotype; 2) sources of variation and functional redundancy within each biological level [96] that exists for any one individual, set against the background of patient and tumour heterogeneity; 3) the ever-increasing need for sufficient bioinformatics expertise, software and hardware to process vast amounts of data generated by these integrative studies; 4) the limited availability of clinical material (e.g. core biopsies) for use in different assays. Furthermore, most systems have not incorporated all levels of complexity; for instance, miRNA data has been largely neglected in most models of data integration.

In addition, such models, in particular in the bottom-up approach, may be superseded by the results of overlaying array CGH data with genome-wide siRNA screening of specific cell-line models [91]. Integrating genomic changes with high-throughput siRNA screening may lead to the identification of amplified genes whose siRNA inhibition is lethal, and may expedite the identification of oncogenes fundamental to the survival of cancer cells.

5.8 Clinical applications

In parallel with and perhaps as a necessary corollary to the search for therapeutic targets, there is now widespread recognition of the essential requirement for biological or biochemical features, i.e. biomarkers, which can be used to measure or predict the effects of treatment. This is particularly important for patient selection, both clinically and economically. given the often considerable financial costs of these drugs.

The availability of companion diagnostic assays for biomarkers that are inexpensive and easily performed, such as the semi-quantitative immunohistochemical assay used to measure the levels of oestrogen receptor (ER) in breast cancer to predict response to anti-oestrogen therapy (as we have seen in Chapter 4), is much more likely to facilitate the translation of any newly-discovered therapeutic agent or strategy into clinical use. Given the potential for the practical and commercial impact of novel drugs to be maximized in this way, companion diagnostics are becoming an integral part of the design and development of targeted cancer therapies, and are likely to play an increasing role in cancer care as we move toward the development of strategies for individualized therapy. In the context of approaches directed towards the identification of candidate oncogenes within recurrently-amplified genomic regions as putative therapeutic targets, the added advantage is that the target doubles up as a biomarker as well. A good example of this is the routine use of fluorescent *in situ* hybridization (FISH) assays in clinical pathology laboratories to detect amplification of the HER2 gene in breast tumours (Chapter 6) [98], which is present in 15–20% of breast cancers and predicts for response to drugs, e.g. trastuzumab, designed to inhibit the effects of the overexpressed protein product. Furthermore, the frequency of the recurrent amplification and the specific tumour subgroup where it is most prevalent can also be established in this manner. By interrogating the genome to identify critical molecular drivers in cancer, aCGH offers the means to identify the therapeutic target and hence the appropriate biomarker assay as well in the process of drug development.

5.9 Conclusions

The promise of aCGH needs to be tempered by the reality that it still remains a crude, albeit powerful, screening tool and should be used as an adjunct to other molecular techniques. The identification of novel amplicons and tumour-suppressor genes [18, 77, 78, 99] has undoubtedly been expedited by aCGH but this is only a small step forward in unravelling the complexity of cancer. Indeed, the way forward is likely to require a systems-biology approach [94, 95], where data from aCGH, gene-expression analysis and high-throughput functional assays (e.g. combining genome-wide aCGH with targeted function analysis of genes pertaining to specific amplicons) is integrated to facilitate our understanding of complex biological systems and serve as a basis for the identification of future therapeutic strategies and targets in cancer. With the development of next-generation sequencers [34–39], it will finally be possible to determine all numerical and structural genetic aberrations in a given tumour in a single experiment, and to quantitatively define all RNA species expressed in the same sample, by means of digital expression profiling and RNA sequencing. These technology developments undoubtedly herald a new era for cancer biology. Understanding the complexity of cancers at the genomic and transcriptomic level, with the resolution and accuracy provided by next-generation sequencers, will constitute a significant step towards the goal of individualized medicine.

Useful Web sites

Illumina (San Diego, USA): http://www.illumina.com and http://www.illumina.com/morethansequencing

Agilent Technologies (Santa Clara, USA): http://www.agilent.com

Roche Nimblegen (Madison, USA): http://www.nimblegen.com

Affmetrix (Santa Clara, USA): http://www.affymetrix.com and http://www.affymetrix.com/technology/mip_technology.affx

The R Project for Statistical Computing: http://www.r-project.org

Bioconductor: http://www.bioconductor.org.

References

[1] Weinstein IB. Cancer: addiction to oncogenes – the Achilles heel of cancer. Science. 2002;297:63–4.
[2] Weinstein IB, Joe AK. Mechanisms of disease: oncogene addiction – a rationale for molecular targeting in cancer therapy. Nat Clin Pract Oncol. 2006;3:448–57.
[3] Hanahan D, Weinberg RA. The hallmarks of cancer. Cell. 2000;100:57–70.
[4] Weinstein IB. Disorders in cell circuitry during multistage carcinogenesis: the role of homeostasis. Carcinogenesis. 2000;21:857–64.
[5] Greaves M. Darwinian medicine: a case for cancer. Nat Rev Cancer. 2007;7:213–21.
[6] Felsher DW, Bishop JM. Reversible tumorigenesis by MYC in hematopoietic lineages. Mol Cell. 1999;4:199–207.
[7] Kantarjian HM, Giles F, Quintas-Cardama A, Cortes J. Important therapeutic targets in chronic myelogenous leukemia. Clin Cancer Res. 2007;13:1089–97.

[8] Lohrisch C, Piccart M. An overview of HER2. Semin Oncol. 2001; 28: 3–11.
[9] Bilancia D, Rosati G, Dinota A, Germano D, Romano R, Manzione L. Lapatinib in breast cancer. Ann Oncol. 2007;18(6):vi26–30.
[10] Hoglund M, Gisselsson D, Sall T, Mitelman F. Coping with complexity: multivariate analysis of tumor karyotypes. Cancer Genet Cytogenet. 2002;135:103–09.
[11] Kallioniemi A, Kallioniemi OP, Sudar D, Rutovitz D, Gray JW, Waldman F, Pinkel D. Comparative genomic hybridization for molecular cytogenetic analysis of solid tumors. Science. 1992;258:818–21.
[12] Tan DS, Lambros MB, Natrajan R, Reis-Filho JS. Getting it right: designing microarray (and not 'microawry') comparative genomic hybridization studies for cancer research. Lab Invest. 2007;87:737–54.
[13] Jong K, Marchiori E, van der Vaart A, Chin SF, Carvalho B, Tijssen M, Eijk PP, van den Ijssel P, Grabsch H, Quirke P, Oudejans JJ, Meijer GA, Caldas C, Ylstra B. Cross-platform array comparative genomic hybridization meta-analysis separates hematopoietic and mesenchymal from epithelial tumors. Oncogene. 2007;26:1499–1506.
[14] Bergamaschi A, Kim YH, Wang P, Sorlie T, Hernandez-Boussard T, Lonning PE, Tibshirani R, Borresen-Dale AL, Pollack JR. Distinct patterns of DNA copy number alteration are associated with different clinicopathological features and gene-expression subtypes of breast cancer. Genes Chromosomes Cancer. 2006;45:1033–40.
[15] Tanami H, Tsuda H, Okabe S, Iwai T, Sugihara K, Imoto I, Inazawa J. Involvement of cyclin D3 in liver metastasis of colorectal cancer, revealed by genome-wide copy-number analysis. Lab Invest. 2005;85:1118–29.
[16] Reis-Filho JS, Simpson PT, Gale T, Lakhani SR. The molecular genetics of breast cancer: the contribution of comparative genomic hybridization. Pathol Res Pract. 2005;201:713–25.
[17] Reis-Filho JS, Simpson PT, Jones C, Steele D, Mackay A, Iravani M, Fenwick K, Valgeirsson H, Lambros M, Ashworth A, Palacios J, Schmitt F, Lakhani SR. Pleomorphic lobular carcinoma of the breast: role of comprehensive molecular pathology in characterization of an entity. J Pathol. 2005;207:1–13.
[18] Cheng KW, Lahad JP, Kuo WL, Lapuk A, Yamada K, Auersperg N, Liu J, Smith-McCune K, Lu KH, Fishman D, Gray JW, Mills GB. The RAB25 small GTPase determines aggressiveness of ovarian and breast cancers. Nat Med. 2004;10:1251–6.
[19] Reis-Filho JS, Simpson PT, Turner NC, Lambros MB, Jones C, Mackay A, Grigoriadis A, Sarrio D, Savage K, Dexter T, Iravani M, Fenwick K, Weber B, Hardisson D, Schmitt FC, Palacios J, Lakhani SR, Ashworth A. FGFR1 emerges as a potential therapeutic target for lobular breast carcinomas. Clin Cancer Res. 2006;12:6652–62.
[20] Reis-Filho JS, Pinheiro C, Lambros MB, Milanezi F, Carvalho S, Savage K, Simpson PT, Jones C, Swift S, Mackay A, Reis RM, Hornick JL, Pereira EM, Baltazar F, Fletcher CD, Ashworth A, Lakhani SR, Schmitt FC. EGFR amplification and lack of activating mutations in metaplastic breast carcinomas. J Pathol. 2006;209:445–53.
[21] Simon R, Radmacher MD, Dobbin K, McShane LM. Pitfalls in the use of DNA microarray data for diagnostic and prognostic classification. J Natl Cancer Inst. 2003;95:14–18.

[22] Solinas-Toldo S, Lampel S, Stilgenbauer S, Nickolenko J, Benner A, Dohner H, Cremer T, Lichter P. Matrix-based comparative genomic hybridization: biochips to screen for genomic imbalances. Genes Chromosomes Cancer. 1997;20: 399–407.
[23] Pollack JR, Perou CM, Alizadeh AA, Eisen MB, Pergamenschikov A, Williams CF, Jeffrey SS, Botstein D, Brown PO. Genome-wide analysis of DNA copy-number changes using cDNA microarrays. Nat Genet. 1999;23:41–46.
[24] Pinkel D, Albertson DG. Array comparative genomic hybridization and its applications in cancer. Nat Genet. 2005;37:S11–17.
[25] Lockwood WW, Chari R, Chi B, Lam WL. Recent advances in array comparative genomic hybridization technologies and their applications in human genetics. Eur J Hum Genet. 2006;14:139–48.
[26] Ylstra B, van den Ijssel P, Carvalho B, Brakenhoff RH, Meijer GA. BAC to the future! or oligonucleotides: a perspective for micro array comparative genomic hybridization (array CGH). Nucleic Acids Res. 2006;34:445–50.
[27] Johnson NA, Hamoudi RA, Ichimura K, Liu L, Pearson DM, Collins VP, Du MQ. Application of array CGH on archival formalin-fixed paraffin-embedded tissues including small numbers of microdissected cells. Lab Invest. 2006;86:968–78.
[28] Little SE, Vuononvirta R, Reis-Filho JS, Natrajan R, Iravani M, Fenwick K, Mackay A, Ashworth A, Pritchard-Jones K, Jones C. Array CGH using whole genome amplification of fresh-frozen and formalin-fixed, paraffin-embedded tumor DNA. Genomics. 2006;87:298–306.
[29] Fan JB, Chee MS, Gunderson KL. Highly parallel genomic assays. Nat Rev Genet. 2006;7:632–44.
[30] Ji H, Kumm J, Zhang M, Farnam K, Salari K, Faham M, Ford JM, Davis RW. Molecular inversion probe analysis of gene copy alterations reveals distinct categories of colorectal carcinoma. Cancer Res. 2006;66:7910–19.
[31] Wang Y, Moorhead M, Karlin-Neumann G, Falkowski M, Chen C, Siddiqui F, Davis RW, Willis TD, Faham M. Allele quantification using molecular inversion probes (MIP). Nucleic Acids Res. 2005;33:e183.
[32] Paterson AL, Pole JC, Blood KA, Garcia MJ, Cooke SL, Teschendorff AE, Wang Y, Chin SF, Ylstra B, Caldas C, Edwards PA. Co-amplification of 8p12 and 11q13 in breast cancers is not the result of a single genomic event. Genes Chromosomes Cancer. 2007;46:427–39.
[33] Bentley DR. Whole-genome re-sequencing. Curr Opin Genet Dev. 2006;16:545–52.
[34] Kahvejian A, Quackenbush J, Thompson JF. What would you do if you could sequence everything?. Nat Biotechnol. 2008;26:1125–33.
[35] von Bubnoff A. Next-generation sequencing: the race is on. Cell. 2008;132:721–3.
[36] Schuster SC. Next-generation sequencing transforms today's biology. Nat Methods. 2008;5:16–18.
[37] Shendure J. The beginning of the end for microarrays?. Nat Methods. 2008;5: 585–7.
[38] Mardis ER. The impact of next-generation sequencing technology on genetics. Trends Genet. 2008;24:133–41.
[39] Campbell PJ, Stephens PJ, Pleasance ED, O'Meara S, Li H, Santarius T, Stebbings LA, Leroy C, Edkins S, Hardy C, Teague JW, Menzies A, Goodhead I, Turner DJ,

Clee CM, Quail MA, Cox A, Brown C, Durbin R, Hurles ME, Edwards PA, Bignell GR, Stratton MR, Futreal PA. Identification of somatically acquired rearrangements in cancer using genome-wide massively parallel paired-end sequencing. Nat Genet. 2008;40:722–9.

[40] van Beers EH, Joosse SA, Ligtenberg MJ, Fles R, Hogervorst FB, Verhoef S, Nederlof PM. A multiplex PCR predictor for aCGH success of FFPE samples. Br J Cancer. 2006;94:333–7.

[41] Baldwin C, Garnis C, Zhang L, Rosin MP, Lam WL. Multiple microalterations detected at high frequency in oral cancer. Cancer Res. 2005;65:7561–7.

[42] Nessling M, Richter K, Schwaenen C, Roerig P, Wrobel G, Wessendorf S, Fritz B, Bentz M, Sinn HP, Radlwimmer B, Lichter P. Candidate genes in breast cancer revealed by microarray-based comparative genomic hybridization of archived tissue. Cancer Res. 2005;65:439–47.

[43] van Dekken H, Paris PL, Albertson DG, Alers JC, Andaya A, Kowbel D, van der Kwast TH, Pinkel D, Schroder FH, Vissers KJ, Wildhagen MF, Collins C. Evaluation of genetic patterns in different tumor areas of intermediate-grade prostatic adenocarcinomas by high-resolution genomic array analysis. Genes Chromosomes Cancer. 2004;39:249–56.

[44] Harvell JD, Kohler S, Zhu S, Hernandez-Boussard T, Pollack JR, van de Rijn M. High-resolution array-based comparative genomic hybridization for distinguishing paraffin-embedded Spitz nevi and melanomas. Diagn Mol Pathol. 2004;13:22–5.

[45] Linn SC, West RB, Pollack JR, Zhu S, Hernandez-Boussard T, Nielsen TO, Rubin BP, Patel R, Goldblum JR, Siegmund D, Botstein D, Brown PO, Gilks CB, van de Rijn M. Gene expression patterns and gene copy-number changes in dermatofibrosarcoma protuberans. Am J Pathol. 2003;163:2383–95.

[46] Thompson ER, Herbert SC, Forrest SM, Campbell IG. Whole genome SNP arrays using DNA derived from formalin-fixed, paraffin-embedded ovarian tumor tissue. Hum Mutat. 2005;26:384–9.

[47] Weiss MM, Hermsen MA, Meijer GA, van Grieken NC, Baak JP, Kuipers EJ, van Diest PJ. Comparative genomic hybridisation. Mol Pathol. 1999;52:243–51.

[48] Chin SF, Teschendorff AE, Marioni JC, Wang Y, Barbosa-Morais NL, Thorne NP, Costa JL, Pinder SE, van de Wiel MA, Green AR, Ellis IO, Porter PL, Tavare S, Brenton JD, Ylstra B, Caldas C. High-resolution aCGH and expression profiling identifies a novel genomic subtype of ER negative breast cancer. Genome Biol. 2007;8:R215.

[49] Arriola E, Lambros MB, Jones C, Dexter T, Mackay A, Tan DS, Tamber N, Fenwick K, Ashworth A, Dowsett M, Reis-Filho JS. Evaluation of Phi29-based whole-genome amplification for microarray-based comparative genomic hybridisation. Lab Invest. 2007;87:75–83.

[50] Raponi M, Zhang Y, Yu J, Chen G, Lee G, Taylor JM, Macdonald J, Thomas D, Moskaluk C, Wang Y, Beer DG. Gene expression signatures for predicting prognosis of squamous cell and adenocarcinomas of the lung. Cancer Res. 2006;66:7466–72.

[51] Pirker C, Raidl M, Steiner E, Elbling L, Holzmann K, Spiegl-Kreinecker S, Aubele M, Grasl-Kraupp B, Marosi C, Micksche M, Berger W. Whole genome

amplification for CGH analysis: linker-adapter PCR as the method of choice for difficult and limited samples. Cytometry A. 2004;61:26–34.

[52] Tanabe C, Aoyagi K, Sakiyama T, Kohno T, Yanagitani N, Akimoto S, Sakamoto M, Sakamoto H, Yokota J, Ohki M, Terada M, Yoshida T, Sasaki H. Evaluation of a whole-genome amplification method based on adaptor-ligation PCR of randomly sheared genomic DNA. Genes Chromosomes Cancer. 2003;38: 168–76.

[53] Fiegler H, Geigl JB, Langer S, Rigler D, Porter K, Unger K, Carter NP, Speicher MR. High resolution array-CGH analysis of single cells. Nucleic Acids Res. 2007;35:e15.

[54] Aviel-Ronen S, Qi Zhu C, Coe BP, Liu N, Watson SK, Lam WL, Tsao MS. Large fragment Bst DNA polymerase for whole genome amplification of DNA from formalin-fixed paraffin-embedded tissues. BMC Genomics. 2006;7:312.

[55] Lovmar L, Syvanen AC. Multiple displacement amplification to create a long-lasting source of DNA for genetic studies. Hum Mutat. 2006;27:603–14.

[56] Lage JM, Leamon JH, Pejovic T, Hamann S, Lacey M, Dillon D, Segraves R, Vossbrinck B, Gonzalez A, Pinkel D, Albertson DG, Costa J, Lizardi PM. Whole genome analysis of genetic alterations in small DNA samples using hyper-branched strand displacement amplification and array-CGH. Genome Res. 2003;13:294–307.

[57] Courjal F, Cuny M, Simony-Lafontaine J, Louason G, Speiser P, Zeillinger R, Rodriguez C, Theillet C. Mapping of DNA amplifications at 15 chromosomal localizations in 1875 breast tumors: definition of phenotypic groups. Cancer Res. 1997;57:4360–7.

[58] Elbauomy Elsheikh S, Green AR, Lambros MB, Turner NC, Grainge MJ, Powe D, Ellis IO, Reis-Filho JS. FGFR1 amplification in breast carcinomas: a chromogenic *in situ* hybridisation analysis. Breast Cancer Res. 2007;9:R23.

[59] Gelsi-Boyer V, Orsetti B, Cervera N, Finetti P, Sircoulomb F, Rouge C, Lasorsa L, Letessier A, Ginestier C, Monville F, Esteyries S, Adelaide J, Esterni B, Henry C, Ethier SP, Bibeau F, Mozziconacci MJ, Charafe-Jauffret E, Jacquemier J, Bertucci F, Birnbaum D, Theillet C, Chaffanet M. Comprehensive profiling of 8p11-12 amplification in breast cancer. Mol Cancer Res. 2005;3:655–67.

[60] Ginestier C, Cervera N, Finetti P, Esteyries S, Esterni B, Adelaide J, Xerri L, Viens P, Jacquemier J, Charafe-Jauffret E, Chaffanet M, Birnbaum D, Bertucci F. Prognosis and gene expression profiling of 20q13-amplified breast cancers. Clin Cancer Res. 2006;12:4533–44.

[61] Natrajan R, Williams RD, Grigoriadis A, Mackay A, Fenwick K, Ashworth A, Dome JS, Grundy PE, Pritchard-Jones K, Jones C. Delineation of a 1Mb breakpoint region at 1p13 in Wilms tumors by fine-tiling oligonucleotide array CGH. Genes Chromosomes Cancer. 2007.

[62] Dobbin K, Simon R. Sample size determination in microarray experiments for class comparison and prognostic classification. Biostatistics. 2005;6:27–38.

[63] Ein-Dor L, Zuk O, Domany E. Thousands of samples are needed to generate a robust gene list for predicting outcome in cancer. Proc Natl Acad Sci USA. 2006;103:5923–8.

[64] Simon R. Roadmap for developing and validating therapeutically relevant genomic classifiers. J Clin Oncol. 2005;23:7332–41.

[65] Pawitan Y, Murthy KR, Michiels S, Ploner A. Bias in the estimation of false discovery rate in microarray studies. Bioinformatics. 2005;21:3865–72.
[66] Paris PL, Andaya A, Fridlyand J, Jain AN, Weinberg V, Kowbel D, Brebner JH, Simko J, Watson JE, Volik S, Albertson DG, Pinkel D, Alers JC, van der Kwast TH, Vissers KJ, Schroder FH, Wildhagen MF, Febbo PG, Chinnaiyan AM, Pienta KJ, Carroll PR, Rubin MA, Collins C, van Dekken H. Whole genome scanning identifies genotypes associated with recurrence and metastasis in prostate tumors. Hum Mol Genet. 2004;13:1303–13.
[67] Beroukhim R, Getz G, Nghiemphu L, Barretina J, Hsueh T, Linhart D, Vivanco I, Lee JC, Huang JH, Alexander S, Du J, Kau T, Thomas RK, Shah K, Soto H, Perner S, Prensner J, Debiasi RM, Demichelis F, Hatton C, Rubin MA, Garraway LA, Nelson SF, Liau L, Mischel PS, Cloughesy TF, Meyerson M, Golub TA, Lander ES, Mellinghoff IK, Sellers WR. Assessing the significance of chromosomal aberrations in cancer: methodology and application to glioma. Proc Natl Acad Sci USA. 104 20 2007 007–12.
[68] Colella S, Yau C, Taylor JM, Mirza G, Butler H, Clouston P, Bassett AS, Seller A, Holmes CC, Ragoussis J. QuantiSNP. An objective Bayes hidden-Markov model to detect and accurately map copy number variation using SNP genotyping data. Nucleic Acids Res. 2007.
[69] Natrajan R, Pinilla S, Marchio C, Williams R, Vatcheva R, Mackay A, Fenwick K, Tamber N, Palacios J, Ashworth A, Reis-Filho J. BAC to INFINIUM: a comparison of near-tiling path BAC arrays and illumina HAP300 SNP arrays for molecular genetic profiling of breast cancer. Glasgow Pathology 2007 Fourth Joint Meeting of the British Division of the International Academy of Pathology and the Pathological Society of Great Britain & Ireland. Abstracts supplement. 2007 13A.
[70] Hupe P, Stransky N, Thiery JP, Radvanyi F, Barillot E. Analysis of array CGH data: from signal ratio to gain and loss of DNA regions. Bioinformatics. 2004;20:3413–22.
[71] Shah SP, Xuan X, DeLeeuw RJ, Khojasteh M, Lam WL, Ng R, Murphy KP. Integrating copy number polymorphisms into array CGH analysis using a robust HMM. Bioinformatics. 2006;22:e431–9.
[72] van Beers EH, Nederlof PM. Array-CGH and breast cancer. Breast Cancer Res. 2006;8:210.
[73] Chari R, Lockwood WW, Lam WL. Computational methods for the analysis of array comparative genomic hybridization. Cancer Inform. 2006;2:48–58.
[74] Ginzinger DG. Gene quantification using real-time quantitative PCR: an emerging technology hits the mainstream. Exp Hematol. 2002;30:503–12.
[75] Di Palma S, Lambros MB, Savage K, Jones C, Mackay A, Dexter T, Iravani M, Fenwick K, Ashworth A, Reis Filho JS. Oncocytic change in pleomorphic adenoma: molecular evidence in support of an origin in neoplastic cells. J Clin Pathol. 2006.
[76] Lambros MB, Simpson PT, Jones C, Natrajan R, Westbury C, Steele D, Savage K, Mackay A, Schmitt FC, Ashworth A, Reis-Filho JS. Unlocking pathology archives for molecular genetic studies: a reliable method to generate probes for chromogenic and fluorescent *in situ* hybridization. Lab Invest. 2006;86: 398–408.

[77] Natrajan R, Reis-Filho JS, Little SE, Messahel B, Brundler MA, Dome JS, Grundy PE, Vujanic GM, Pritchard-Jones K, Jones C. Blastemal expression of type I insulin-like growth factor receptor in Wilms' tumors is driven by increased copy number and correlates with relapse. Cancer Res. 6611 2006 148–55.

[78] Natrajan R, Little SE, Reis-Filho JS, Hing L, Messahel B, Grundy PE, Dome JS, Schneider T, Vujanic GM, Pritchard-Jones K, Jones C. Amplification and overexpression of CACNA1E correlates with relapse in favorable histology Wilms' tumors. Clin Cancer Res. 2006;12:7284–93.

[79] Vincent-Salomon A, Gruel N, Lucchesi C, Mac Grogan G, Dendale R, Sigal-Zafrani B, Longy M, Raynal V, Pierron G, de Mascarel I, Taris C, Stoppa-Lyonnet D, Pierga JY, Salmon R, Sastre-Garau X, Fourquet A, Delattre O, de Cremoux P, Aurias A. Identification of typical medullary breast carcinoma as a genomic sub-group of basal-like carcinomas, a heterogeneous new molecular entity. Breast Cancer Res. 2007;9:R24.

[80] Carter NP. Methods and strategies for analyzing copy number variation using DNA microarrays. Nat Genet. 2007;39:S16–S21.

[81] Loo LW, Grove DI, Williams EM, Neal CL, Cousens LA, Schubert EL, Holcomb IN, Massa HF, Glogovac J, Li CI, Malone KE, Daling JR, Delrow JJ, Trask BJ, Hsu L, Porter PL. Array comparative genomic hybridization analysis of genomic alterations in breast cancer subtypes. Cancer Res. 2004;64:8541–9.

[82] Perou CM, Sorlie T, Eisen MB, van de Rijn M, Jeffrey SS, Rees CA, Pollack JR, Ross DT, Johnsen H, Akslen LA, Fluge O, Pergamenschikov A, Williams C, Zhu SX, Lonning PE, Borresen-Dale AL, Brown PO, Botstein D. Molecular portraits of human breast tumours. Nature. 2000;406:747–52.

[83] Pierga JY, Reis-Filho JS, Cleator SJ, Dexter T, Mackay A, Simpson P, Fenwick K, Iravani M, Salter J, Hills M, Jones C, Ashworth A, Smith IE, Powles T, Dowsett M. Microarray-based comparative genomic hybridisation of breast cancer patients receiving neoadjuvant chemotherapy. Br J Cancer. 2007;96:341–51.

[84] Stange DE, Radlwimmer B, Schubert F, Traub F, Pich A, Toedt G, Mendrzyk F, Lehmann U, Eils R, Kreipe H, Lichter P. High-resolution genomic profiling reveals association of chromosomal aberrations on 1q and 16p with histologic and genetic subgroups of invasive breast cancer. Clin Cancer Res. 2006;12:345–52.

[85] van de Vijver MJ, He YD, van't Veer LJ, Dai H, Hart AA, Voskuil DW, Schreiber GJ, Peterse JL, Roberts C, Marton MJ, Parrish M, Atsma D, Witteveen A, Glas A, Delahaye L, van der Velde T, Bartelink H, Rodenhuis S, Rutgers ET, Friend SH, Bernards R. A gene-expression signature as a predictor of survival in breast cancer. N Engl J Med. 2002;347:1999–2009.

[86] Garcia MJ, Pole JC, Chin SF, Teschendorff A, Naderi A, Ozdag H, Vias M, Kranjac T, Subkhankulova T, Paish C, Ellis I, Brenton JD, Edwards PA, Caldas C. A 1 Mb minimal amplicon at 8p11-12 in breast cancer identifies new candidate oncogenes. Oncogene. 2005;24:5235–45.

[87] Neve RM, Chin K, Fridlyand J, Yeh J, Baehner FL, Fevr T, Clark L, Bayani N, Coppe JP, Tong F, Speed T, Spellman PT, DeVries S, Lapuk A, Wang NJ, Kuo WL, Stilwell JL, Pinkel D, Albertson DG, Waldman FM, McCormick F, Dickson RB, Johnson MD, Lippman M, Ethier S, Gazdar A, Gray JW. A collection of breast cancer cell lines for the study of functionally distinct cancer subtypes. Cancer Cell. 2006;10:515–27.

[88] Jonsson G, Staaf J, Olsson E, Heidenblad M, Vallon-Christersson J, Osoegawa K, de Jong P, Oredsson S, Ringner M, Hoglund M, Borg A. High-resolution genomic profiles of breast cancer cell lines assessed by tiling BAC array comparative genomic hybridization. Genes Chromosomes Cancer. 2007;46:543–58.

[89] Greshock J, Nathanson K, Martin AM, Zhang L, Coukos G, Weber BL, Zaks TZ. Cancer cell lines as genetic models of their parent histology: analyses based on array comparative genomic hybridization. Cancer Res. 2007;67:3594–600.

[90] Feber A, Clark J, Goodwin G, Dodson AR, Smith PH, Fletcher A, Edwards S, Flohr P, Falconer A, Roe T, Kovacs G, Dennis N, Fisher C, Wooster R, Huddart R, Foster CS, Cooper CS. Amplification and overexpression of E2F3 in human bladder cancer. Oncogene. 2004;23:1627–30.

[91] Iorns E, Lord CJ, Turner N, Ashworth A. Utilizing RNA interference to enhance cancer drug discovery. Nat Rev Drug Discov. 2007;6:556–68.

[92] Kao J, Pollack JR. RNA interference-based functional dissection of the 17q12 amplicon in breast cancer reveals contribution of coamplified genes. Genes Chromosomes Cancer. 2006;45:761–9.

[93] Guan Y, Kuo WL, Stilwell JL, Takano H, Lapuk AV, Fridlyand J, Mao JH, Yu M, Miller MA, Santos JL, Kalloger SE, Carlson JW, Ginzinger DG, Celniker SE, Mills GB, Huntsman DG, Gray JW. Amplification of PVT1 contributes to the pathophysiology of ovarian and breast cancer. Clin Cancer Res. 2007;13:5745–55.

[94] Bosl WJ. Systems biology by the rules: hybrid intelligent systems for pathway modeling and discovery. BMC Syst Biol. 2007;1:13.

[95] Hornberg JJ, Bruggeman FJ, Westerhoff HV, Lankelma J. Cancer: a systems biology disease. Biosystems. 2006;83:81–90.

[96] Reif DM, White BC, Moore JH. Integrated analysis of genetic, genomic and proteomic data. Expert Rev Proteomics. 2004;1:67–75.

[97] Greenbaum D, Jansen R, Gerstein M. Analysis of mRNA expression and protein abundance data: an approach for the comparison of the enrichment of features in the cellular population of proteins and transcripts. Bioinformatics. 2002;18: 585–96.

[98] Lambros MB, Natrajan R, Reis-Filho JS. Chromogenic and fluorescent *in situ* hybridization in breast cancer. Hum Pathol. 2007;38:1105–22.

[99] Rivera MN, Kim WJ, Wells J, Driscoll DR, Brannigan BW, Han M, Kim JC, Feinberg AP, Gerald WL, Vargas SO, Chin L, Iafrate AJ, Bell DW, Haber DA. An X chromosome gene, WTX, is commonly inactivated in Wilms tumor. Science. 2007;315:642–5.

6

Tissue *In Situ* Hybridization

Anthony O' Grady[1], John O' Loughlin[2] and Hilary Magee[3]

[1]*Chief Medical Scientist, Department of Pathology, Royal College of Surgeons in Ireland Education & Research Centre, Beaumont Hospital*
[2]*Chief Medical Scientist, Department of Cellular Pathology, Adelaide & Meath Hospital Dublin, incorporating the National Children's Hospital*
[3]*Senior Medical Scientist, Department of Cellular Pathology, Adelaide & Meath Hospital, Dublin, incorporating the National Children's Hospital*

6.1 Introduction

The technique of DNA–DNA hybridization was first described in 1961 [1]; however it is Gall and Pardue [2] and John *et al.* [3] who are credited with developing the technique we are familiar with today. The initial experiments focused on the identification of specific DNA targets using RNA or DNA labelled with radioisotopes such as tritium (^{3}H). The intrinsic problems associated with the use of these and other radioactively-labelled probes meant that initially, widespread application of the technique was restricted. It was not until the introduction of nonradioactive labels [4], combined with novel methods of probe production, that there was an exponential expansion in the application and use of the *in situ* hybridization (ISH) technique.

Today, fluorescent *in situ* hybridization (FISH) has provided significant advances in resolution, speed and safety, and has paved the way for the development of simultaneous detection of multiple targets. The use of interphase FISH to study cytogenetic abnormalities in routinely formalin-fixed paraffin-embedded (FFPE) tissue has become commonplace. Tissue ISH has become an invaluable diagnostic and research tool in applications requiring morphological localization of nucleic acid sequences. It is widely

Advanced Techniques in Diagnostic Cellular Pathology Edited by Mary Hannon–Fletcher and Perry Maxwell

used to detect DNA, RNA and more recently microRNA in tissue sections and cytological preparations. Tissue ISH is used to detect nucleic acid sequences in a wide variety of solid neoplastic and infectious conditions and is becoming a crucial theranostic tool, helping to guide therapy by identifying relevant targets for new drugs in the field of pharmacogenomics.

New and improved methodologies are constantly being described, including the use of commercially-available labelled probes (fluorochrome, biotin and digoxigenin); silver (SISH) and quantum dot (Q dot) detection systems; and dual ICC and FISH (FICTION) and dual-colour chromogenic (CISH) enhancement products. Moreover, the time-consuming and complex procedures involved in tissue ISH have been simplified by the introduction of automated slide pretreatment and hybridization stations. These semi/fully-automated systems ensure the reliability, reproducibility and standardization of ISH procedures, increase assay throughput and improve procedure turnaround times.

The focus of this chapter will be on the basics of the ISH method, its application to tissue sections in particular, its use in diagnostic and prognostic pathology, the advances that have taken place since it was introduced and novel developments associated with it.

6.2 ISH probes

The basic requirements of a probe for use in tissue ISH are that it is complementary to the target nucleic acid sequence and can be labelled in such a way as to allow microscopic visualization of the hybrid formed. The different types of probe include the following.

6.2.1 DNA probes

Double-stranded DNA probes are most commonly used to detect DNA targets. They are generated by the cloning and amplification of specific sequences of DNA or cDNA, derived by reverse transcription of mRNA employing vectors (bacterial plasmids and cosmids). The resulting sequence can be labelled by nick translation or random priming. Using whole plasmid DNA can increase the intensity of the hybridization signal due to the formation of multiple hybrids between plasmid sequences at the target site, but may also increase the level of background staining [5]. On the other hand, the use of the cloned insert alone can produce a cleaner signal but reduces the sensitivity of the hybridization reaction. Prior to hybridization, a double-stranded DNA probe and its target must be denatured by physical (heat) or chemical (formamide) means to allow annealing.

Single-stranded DNA probes can be generated using the polymerase chain reaction (PCR). Using a DNA template and the appropriate primers, probes with high specificity can be produced. An advantage of this method of probe generation is the ability to incorporate labelling of the probe into the PCR reaction.

6.2.2 RNA probes

Single-stranded RNA probes, so-called riboprobes, are most commonly used to detect RNA in tissue sections. Riboprobes are usually generated by *in vitro* transcription from plasmids containing the sequence of interest. The plasmids are designed with promoter sites for RNA polymerases (e.g. T3, T7 and SP6) and can produce probes complementary to the target RNA sequence (antisense) or identical to the target sequence (sense). The latter is used as the negative control when performing RNA ISH. An alternative method of riboprobe generation is to utilize PCR with appropriate primer sets incorporating RNA polymerase promoter sequences at their 5′ ends [6]. This method is less laborious and relatively less expensive than cloning. The transcription reaction used to generate riboprobes incorporates labelled uridine instead of thymidine. Although they produce extremely stable hybrids with RNA or DNA, these probes are highly susceptible to degradation by RNases. They can also bind nonspecifically, resulting in high background noise [7].

6.2.3 Oligodeoxynucleotide probes

These are short sequences of DNA, usually 15–50 bases, generated on an automated DNA synthesizer. Their short length makes them ideal for accessing targets within FFPE tissue sections but they are more likely to be dislodged if excessive post-hybridization stringency washes are used. In addition, their short length may result in nonspecific binding and an increase in background staining. Labelling of oligodeoxynucleotide probes (oligomers) is usually achieved by 3′ end-labelling with deoxynucleotide transferase or by the addition of a label at the 5′ end [8]. The level of sensitivity of oligodeoxynucleotide probes is lower than that achieved with double-stranded DNA probes but this can be improved by using a cocktail of oligomers complementary to multiple adjacent sequences in the target gene/region [9].

6.2.4 Padlock probes

Padlock probes were first described by Nilsson *et al.* [10] in an attempt to prevent mismatch hybridization of single bases. These short oligonucleotide probes (approximately 20 nucleotides long at the 3′ and 5′ end) have a 30–40 mer spacer arm between the 3′ and 5′ ends. During hybridization, a ligation step covalently links the probe sequences to close the padlock probe, forming a circular molecule almost impervious to high-stringency washes. Haptens such as biotin or digoxigenin are attached to the spacer arm for detection purposes.

6.2.5 PNA probes

PNA (polypeptide nucleic acid) is a DNA analogue that forms very stable duplexes with complementary DNA or RNA sequences. The duplexes formed with these probes possess high thermal stability and the probes hybridize with a high affinity and specificity because of their uncharged and flexible polyamide backbone. For a review of PNA applications, see [11].

6.3 Probe labels

Early ISH techniques utilized DNA and RNA probes radioactively labelled with ^{32}P, ^{35}S, ^{125}I and ^{3}H because of their high sensitivity and the amplification effect provided by autoradiographic development. So-called isotopic ISH, although sensitive to the single-copy level, has signal development (by means of dipping slides in liquid photographic emulsion), requiring days or weeks to produce the required result. In addition, identifying the precise localization of the target nucleic acid sequence can be hampered by the spread of the radioactive signal. Due to safety and stability issues, as well as the speed and the quality of target visualization, radioactively-labelled probes have been gradually replaced, such that they are now confined to the research setting.

Biotinylated probes were among the first non-isotopic probes to be used for tissue ISH. They have been widely used for the detection of both DNA and RNA sequences in tissue samples. The affinity of biotin to avidin and its derivatives (e.g. streptavidin) have made it possible to detect relatively low-abundance sequences through the amplification effect provided by the conjugation of many reporter molecules (alkaline phosphatase, horseradish peroxidase, fluorochromes, colloidal silver and gold) on to each avidin molecule. A significant disadvantage when using biotinylated probes is the level of background staining associated with endogenous biotin, particularly in kidney and liver. Techniques for blocking endogenous biotin are available but produce variable results.

Digoxigenin-labelled probes were first used in the late 1980s to detect HPV sequences in cultured cells [12]. They have since gained widespread use in brightfield ISH applications. The advantages over biotin include increased sensitivity and reduced background staining, as digoxigenin is not naturally present in human tissue [13]. Digoxigenin can be incorporated into probes by a number of methods, including random priming, PCR, nick translation, 3′-end labelling and *in vitro* transcription, making it suitable for the generation of cDNA, cRNA and oligonucleotide DNA probes. These probes can then be visualized in the tissue section using anti-digoxigenin antibodies conjugated to alkaline phosphatase or horseradish peroxidise, with appropriate substrates such as 3,3′-diaminobenzidine tetrahydrochloride (DAB) or 3-amino-9-ethylcarbazole (AEC), to give brown or red end colours, respectively. To improve the sensitivity of the reaction an unconjugated anti-digoxigenin antibody may be followed by a conjugated secondary antibody.

At the same time as digoxigenin-labelled probes were being developed, fluorescently-labelled probes were gaining widespread use [14]. A variety of fluorochromes are now available for labelling ISH probes, and advances in FFPE tissue digestion and pretreatment methods have made these the most popular probe type in use in the clinical histopathology laboratory. Fluorescently-labelled probes and antibodies can be used in direct or indirect ISH methods either singly or in combination with multiple other probe colours. Most fluorescently-labelled probes can be visualized directly using a fluorescent microscope with appropriate filters. Background staining is usually low if pretreatment methods are optimized, and FISH has the advantage of speed over other ISH methods, which require the addition of secondary antibodies and subsequent chromogenic detection.

Q dots are also gaining favour and use as fluorochromes in ISH techniques. These are fluorescent semiconductor nanocrystals which possess a number of unique characteristics. Due to the fact that Q dots have broad absorption bands and narrow emission bands, dots of varying sizes can be used to tag numerous different targets simultaneously within the same tissue section. These can then be excited with a single light source of short wavelength, giving rise to emission spectra of varying wavelengths, and therefore colours, allowing distinct signal detection. By using an excitation light source wavelength several hundred nanometres shorter than that of the emission wavelength, the problems associated with intrinsic fluorochrome autofluorescence are minimized. Q dots are also resistant to photobleaching and exhibit a photostability 1000 times greater than conventional fluorochromes such as fluorescein. For a comprehensive review of this area, see [15].

More recently the development of dual-colour chromogenic detection methods has expanded the diagnostic applications of tissue ISH, making it possible to locate more than one gene in a single section without the need for sophisticated fluorescent microscopy facilities.

6.4 ISH detection systems

In situ hybridization using fluorescently-labelled probes does not require a detection system and the reaction can be viewed under the fluorescent microscope with appropriate filters. Immunocytochemistry detection systems are used for the visualization of all other nonradioactive probes. Systems based on alkaline phosphatase-labelled antibodies with nitro-blue tetrazolium, 5-bromo-4-chloro-3-indolyl phosphate (NBT/BCIP) as the chromogen produce a blue/black colour at the site of hybridization. The requirement for aqueous mounting of such sections is a disadvantage. The use of horseradish peroxidase-labelled antibodies with DAB as the chromogen produces a permanent brown colour. Although fluorescently-labelled probes do not require further detection, anti-fluorochrome antibodies can be used to convert the fluorescent label to an immunohistochemically-detectable tag, making it possible to visualize the hybridization site under brightfield microscopy.

6.5 Tissue preparation and the ISH procedure

The *in situ* hybridization technique can be applied to a variety of cellular materials, from cell suspensions and touch preparations (Chapter 2) to tissue sections. Optimization of fixation and processing protocols for these samples is essential to ensuring consistency and validity of results. The use of cell suspensions, touch preparations or fresh-frozen sections produces the best results as there is less degradation of the target sequences of interest compared with fixed tissue. However, quite often the only material available for study is fixed and paraffin-wax embedded. The advantage of using this material is that numerous ISH experiments using a variety of probes can be performed on sequential sections from the same tissue block. Studies can also be carried out retrospectively on archived material. The disadvantage of using this material is the degradation of the target sequence, in particular RNA targets. Therefore tissue samples should be fixed as rapidly as possible post-mortem, or when the tissue sample is obtained.

The fixative of choice in most laboratories is 10% formalin. Formalin is a 37% aqueous solution of the gas formaldehyde. It acts on protein residues and traps DNA and RNA targets in a meshwork of cross-linked methylene bridges (Chapter 4). For a comprehensive review of the fixation effects of various fixatives on DNA and RNA, see [16]. Other fixatives, such as Bouins, are less commonly used and should be avoided for ISH applications as preservation of nucleic acids is suboptimal.

The optimum fixation time for tissues trimmed to processing dimensions is thought to be 24–48 hours. Successful hybridization experiments have been achieved with tissue samples fixed for considerably longer time periods than this, but modifications to the standard technique (digestion times and proteolytic enzyme concentration) are usually required. Prolonged fixation times result in extensive protein cross-linking, making the target nucleic acid less accessible.

Following fixation, the tissue sample can be processed to paraffin wax through a graded series of alcohols and clearing agents. This procedure is automated in most laboratories and results in a tissue sample infiltrated with molten wax, which is subsequently embedded in a mould and the wax allowed to solidify. Once processed the optimum thickness for tissue sections for ISH procedures is approximately four microns. Some truncation of nuclei will be evident in sections of this thickness but this usually does not cause problems in the interpretation of results. Due to the harsh nature of the procedure (use of heat, alkali/acids, detergents, proteolytic enzymes, etc.), detachment of sections from the glass slide can present a problem. This can be largely overcome by the use of electrostatically-charged slides, which can be purchased commercially or produced in-house by coating the slides with poly-L-lysine or 3-amino-propyl-triethoxy-silane.

After sectioning and removal of wax from the tissue it must be treated to unmask the target DNA/RNA. Proteolytic enzymes such as proteinase K, pepsin and pronase are most commonly used to carry out this task. Optimization of time, concentration and temperature is usually required, depending on the amount of target present in the section and the length of fixation of the tissue sample prior to processing. The corollary to this however is that over-digestion can lead to loss of target, so this step in particular must be strictly controlled. The degree of digestion can be checked during the procedure by staining the tissue with DAPI and examining it under the fluorescent microscope for evidence of over- or under-digestion.

After unmasking the target sequence it must be denatured (if double-stranded) prior to hybridization. This usually involves exposing the tissue to heat or a strong base such as sodium hydroxide. The probe, if double-stranded, must also be denatured. Following separate denaturation of target and probe, the binding of probe to target occurs due to the interaction of complimentary base pairs; guanine (G) to cytosine (C) and adenine (A) to thymidine (T), or uracil when the target is RNA. Factors which affect the strength of the bond between the target and the probe include the presence of monovalent cations, organic solvents, the G/C content and heat. Monovalent cations (supplied by standard sodium citrate (SSC)) are positively-charged ions that help to reduce the repulsive forces associated with the negatively-charged nucleotides. Organic solvents such as formamide act by destabilizing hydrogen bonds, thereby allowing hybrid separation and reannealing to occur at lower temperatures. Due to the presence of three hydrogen bonds in the G/C complex, as opposed to two in the A/T complex, the higher the concentration of G/C the more stable the hybrid and the less likely it is to dissociate.

Following hybridization, which for convenience is usually carried out overnight (at 37–55 °C), stringency washes are necessary to remove any noncomplementary strands that may have loosely bound to one another. Increased stringency is achieved by decreasing the salt concentration and/or increasing the temperature of the stringency wash solution. Detection or visualization of the bound probe is dependent on the reporter molecule (isotopic or non-isotopic) linked to the probe and is described above.

6.6 Signal amplification

The problems associated with the lack of sensitivity of non-isotopic versus isotopic ISH were addressed to a certain extent when signal amplification techniques were developed. The technique is based on the catalysed deposition of labelled tyramine by horseradish peroxidase, at the site of probe binding. Tyramide signal amplification (TSA), catalysed signal amplification (CSA) and catalysed reporter deposition signal amplification (CARD) are all synonymous and based on the amplification method first described by Bobrow [17]. Biotinylated tyramine is deposited on to electron-rich moieties such as tyrosine and tryptophan at or near the site of bound horseradish peroxidise-labelled probe. The biotin on the tyramine acts as future binding site for streptavidin–biotin complexes or enzyme- and fluorochrome-labelled streptavidin. This amplification step results in an approximate 100-fold increase in sensitivity, leading to detection of targets of low copy number in cells. Fluorochrome-labelled tyramide amplification can also be used for fluorescent-based methods, and tyramide amplification combined with the use of Q dots has also been reported [18].

The indiscriminate amplification of signal contributes to one of the disadvantages associated with this technique. The method does not distinguish true target/probe binding from nonspecifically-bound probe and therefore, if not removed with post-hybridization stringency washes, the nonspecifically-bound probe will be amplified, leading to an increase in signal-to-noise ratio (background). Endogenous peroxidases within the tissue, if not removed, may also catalyse the deposition of the labelled tyramide, resulting in nonspecific signal amplification. All things considered however, amplification techniques present an efficient and highly-sensitive method for low-copy-number nucleic acid target detection, and they have been used in viral particle detection, combined ICC and ISH and multi-target FISH.

6.7 ISH controls

The use of tissue and probe controls is essential to guarantee the sensitivity and specificity of the ISH reaction. The inclusion of appropriate controls in the experimental procedure provides evidence that the probe is binding selectively to the target and not to other closely-related sequences within the tissue. It can also highlight problems with the experimental protocol or preservation of nucleic acid sequences within the tissue itself.

All ISH experiments should include, where possible, a section from a piece of tissue known to contain the target sequence of interest (positive tissue control). This tissue should ideally be fixed and processed in a similar manner to the tissue of interest. If there is no staining present, this suggests there is a problem with the probe or ISH procedure. The use of probes to housekeeping genes, such as actin, should also be included in the experiment to ensure the efficacy of the procedure and the preservation of nucleic acid sequences within the tissue (positive probe control). In mRNA ISH experiments a poly(dT) probe to detect total mRNA poly A tails should be used. A weak or absent signal with this probe suggests poor preservation of mRNA within the tissue. An effective method to test the specificity of the hybridization reaction is to digest the tissue with DNases or RNases prior to incubation with the probe of interest (negative tissue control). The absence of staining in the digested versus undigested section shows the specificity of the binding reaction. The use of 'nonsense' probes, or in the case of mRNA ISH, sense probes, can give a measure of the extent of nonspecific binding within the tissue of interest (negative probe control). In any ISH experiment the quality of the result obtained is only as good as the quality of the controls (positive and negative) used.

Another form of quality control exists for clinical diagnostic laboratories performing ISH testing. External quality assessment enables unbiased monitoring of the performance of laboratories in the detection of specific diagnostic markers. The United Kingdom National External Quality Assurance Scheme for Immunocytochemistry and *In Situ* Hybridization (UK NEQAS ICC & ISH) organization (see Chapter 4)was set up in 2004 to audit laboratories' analytical and interpretative performances for the HER-2 assay. Three times per year, UK NEQAS ICC & ISH circulates unstained sections prepared from a block containing four control cell lines with different levels of HER-2 gene expression. Participating laboratories are asked to stain and score the sections with their in-house ISH method. Scores are returned to UK NEQAS ICC & ISH and the results are measured against those obtained by a group of eight reference laboratories. Participating laboratories are thus able to assess whether their results are comparable with those of others. On a quarterly basis a journal is circulated to participating laboratories, offering information on best methods for performing the assay. External quality assurance programmes such as this should be a key component of any centre providing a diagnostic ISH service.

6.8 Clinical applications of tissue ISH

The deciphering of the human genome, in combination with recent developments in nucleic acid-based testing, has positioned tissue ISH as a central tool in diagnosis and predictive therapy. Molecular diagnostics including tissue ISH are widely used in the areas of inherited genetic disorders and infectious diseases, as well as haematologic and solid tumours. In addition, tissue ISH has a role to play in guiding appropriate therapy.

6.8.1 Solid tumours

The generally accepted model of tumourigenesis is one of genetic and epigenetic alterations that activate oncogenes and inactivate tumour-suppressor genes. Most solid tumours exhibit a marked degree of chromosomal instability, and different cells within a tumour can exhibit significant chromosomal variation. Chromosomal alterations identified in tumours include aneuploidy (abnormal number of chromosomes), deletion, amplification and translocation. These alterations can be detected at the cellular level by FISH. There are two types of FISH probe used in the analysis of tumours: chromosome enumeration probes (CEPs) and locus-specific indicator (LSI) probes.

The diagnosis of haematologic malignancies has benefited greatly from molecular diagnostic testing in the last 30 years but it is only recently that the application of these tests has started to have an impact on the diagnosis

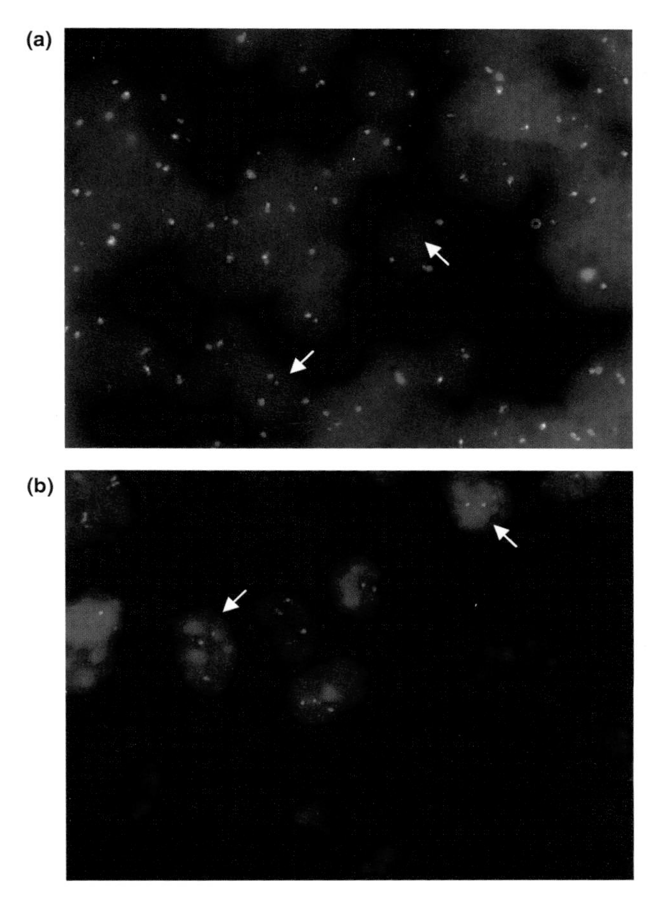

Figure 6.1 The application of FISH to lymphoma and central nervous system tumour diagnosis: (a) mantle cell lymphoma showing the characteristic t(11;14). A breakapart probe identifies cells with one fusion signal (overlapping red and green) and two separate red and green signals (arrows), indicating a chromosomal breakpoint; (b) an oligodendroglioma showing cells with a 1p36 (red) deletion (arrows)

of solid tumours. Interphase FISH is used to detect t(11;14), characteristic of mantle cell lymphomas (Figure 6.1a), producing a fusion of the CCND1 gene and the immunoglobulin-heavy chain locus. This results in the overexpression of the cyclin D1 protein, which can be detected by ICC. Thus serial sections from the same FFPE sample can be used for morphological assessment by H&E, ICC analysis for cyclin D1 protein and FISH analysis for the characteristic translocation, to provide a definitive diagnosis of this particular tumour. Another useful application of tissue ISH is the detection of kappa and lambda mRNA to establish evidence of monoclonality in a suspected case of lymphoma.

As with other subspecialties of pathology, tissue ISH is beginning to play an important role in surgical neuropathology. Unlike many haematologic and soft-tissue tumours, brain tumours are not characterized by signature translocations and fusion products. The main genetic alterations include chromosomal/gene gains/losses. An example of a FISH assay in use in neuropathology is the detection of the 1p (Figure 6.1b) and 19q deletions for prognosis and therapeutic guidance in oligodendrogliomas [19]. These tumours behave less aggressively than astrocytomas, with a slower progression and subsequent longer patient survival. The response to therapies such as PCV (procarbazine, CCNU, vincristine) makes it important to identify this tumour type. FISH studies have shown 1p and 19q co-deletions in up to 90% of oligodendrogliomas [20].

Amplification of the HER-2 gene occurs in more than 20% of breast carcinomas and is associated with a poor outcome [21]. With the development of an effective humanized monoclonal antibody treatment for breast carcinomas overexpressing the HER-2 protein it has become necessary to assess the gene and protein status to identify those patients that will benefit from this. The evaluation of HER-2 gene amplification in breast carcinoma is the best example of a molecular assay used to guide therapy. Cases for HER-2 FISH analysis are chosen on the basis of the immunohistochemical assessment of the HER-2 protein. Equivocal cases (those scoring 2+ with the HercepTest method) are usually referred for confirmatory FISH testing. The most common form of this assay uses a combination of a centromeric chromosome 17 probe labelled with one colour (usually fluorescein/spectrum green) and an allele-specific probe for the HER-2 gene labelled with a second colour (usually Texas red/spectrum orange). A normal non-amplified cell is identified as having two green and two red signals (a ratio of 1), whereas a breast cancer cell displaying HER-2 gene amplification shows four or more red and only two green signals (a ratio of 2 or more). The rationale for FISH testing in 2+ breast cancer cases is that only a proportion (17–25%) of these will demonstrate HER-2 gene amplification and likely benefit from the targeted treatment (Herceptin). The advantage of HER-2 FISH over HER-2 ICC in assessing borderline cases lies in the objectivity of a simple coloured dot-counting method compared to a subjective measure of cell membrane-staining extent and intensity (Chapter 4). In recent years a number of alternative techniques, including SISH and dual-colour chromogenic ISH (Duo-CISH), have been developed to detect HER-2 gene amplification (Figure 6.2).

The identification of characteristic gene rearrangements associated with many soft-tissue tumours has led to the development of diagnostic probe sets that can be used on FFPE material. These tumours can be diagnosed with molecular methodologies such as PCR and RT-PCR, but more and

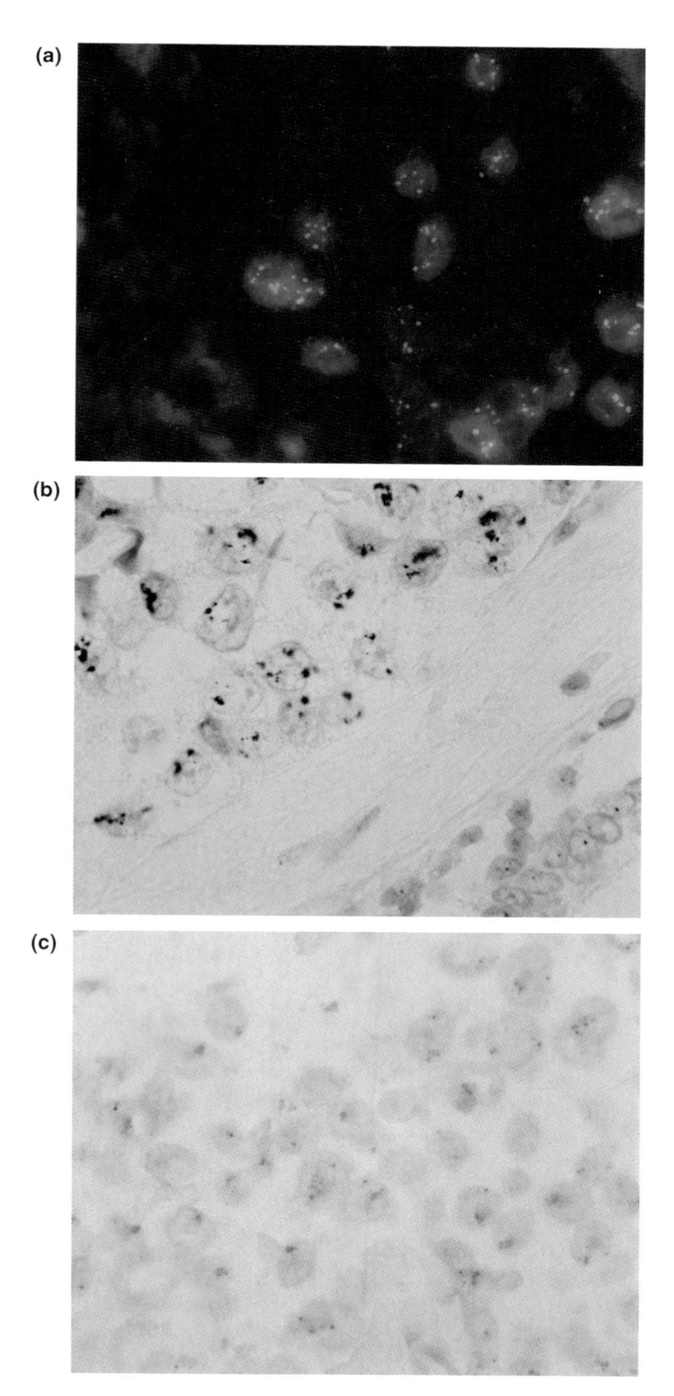

Figure 6.2 ISH detection of HER-2 gene amplification in breast cancer: (a) FISH showing chromosome 17 centromere (green) and multiple copies of the HER-2 gene (red); (b) SISH showing HER-2 gene in black; (c) Duo-CISH showing chromosome 17 centromere (blue) and HER-2 gene (red)

more laboratories are using FISH as the first-line diagnostic test as it has the advantage of preserving the morphology of the sample. For example, the detection of the EWS/FLI1 fusion gene produced by t(11;22)(q24;q12) is used to diagnose peripheral neuroectodermal/Ewing sarcoma [22].

The detection and treatment of bladder cancer is challenging and the prognosis for the patient in terms of disease-free survival, morbidity and quality-of-life issues depends greatly on diagnosing the tumour before it spreads into the muscle wall. Cystoscopy and urine cytology are the traditional diagnostic and screening tools used for this disease, but they exhibit variable sensitivity and specificity. In 2000, a multicolour FISH probe set for the detection of bladder cancer in voided urine was developed [23]. The probe set (UroVysion-Abbott Molecular Inc, Des Plaines, USA) contains CEP probes for chromosomes 3, 7 and 17 (labelled with red, green and aqua, respectively), and an LSI probe (labelled with gold) to the 9p21 region, which together detect the most common aneuploidies occurring in bladder cancer. A scoring system was devised, with a positive result recorded when $\geq$4% of 100 counted cells showed at least four signals of any CEP probe or $\geq$12% of 100 counted cells showed one signal or less of probe 9p21 [24]. This four-target, multicolour FISH approach has resulted in a sensitivity of 92%, compared to 76% for cytology on the same specimens. FISH analysis of cells isolated from bladder washings or voided urine also holds promise for predicting recurrence and progression of the disease and monitoring treatment outcomes.

Table 6.1 highlights examples of solid tumours where tissue ISH has a role to play in diagnosis/predictive therapy.

6.8.2 Infectious agents

DNA and RNA ISH are very useful for the detection of infectious agents in tissue sections. ISH methods can be used to detect bacteria including *Helicobacter pylori* and *Mycoplasma pneumoniae* [25,26]. However, ISH is more frequently used to detect viral infections associated with human diseases such as human immunodeficiency virus (HIV), cytomegalovirus (CMV), human papillomavirus (HPV; Chapter 2), herpes simplex virus (HSV), hepatitis B virus (HBV) and Epstein-Barr virus (EBV). HPV probes specific for low- and high-risk subtypes are now widely available to identify the HPV types associated with neoplastic development (Figure 6.3a) [27]. EBV infections are found to be associated with tumours including Burkitt lymphoma, Hodgkin disease and nasopharyngeal carcinoma. RNA ISH with oligonucleotide or riboprobes has proven successful in detecting latent

Table 6.1 Examples of chromosomal and genetic targets for ISH testing of solid tumours

Tumour Type	*Genetic Signature*	*Reference*
Lymphoma		[40]
Mantle cell	t(11;14)(q13;q32)	
Follicular	t(14;18)(q32;q21) and +3	
MALT	t(11;18)(q21;q21), +3, +7, +12 and +18	
Sarcoma		[41]
Ewing sarcoma/PNET	t(11;22)(q24;q12)	
	t(21;22)(q22;q12)	
	t(7;22)(p22;q12)	
	t(2;22)(q33;q12)	
	t(17;22)(q12;q12)	
	inv(22)	
Synovial sarcoma	t(X;18)(p11;q11)	
Clear-cell sarcoma	t(12;22)(q13;q12)	
Alveolar rhabdomyosarcoma	t(2;13)(q35;q14)	
	t(1;13)(p36;q14)	
Central Nervous System		[42]
Diffuse astrocytomas	7, 9, 10q, 19q	
Oligodendroglial tumours	1p, 9p, 10q, 19q	
Ependymomas	1q, 6q, 18, 22q	
Embryonal tumours	17, 10q, 22	
Meningiomas	1p, 14q, 17q23, 18, 22q, NF2	
Prostate Cancer	Overexpression of AMACR and Hepsin Mutations of KLF-6 and PTEN	[43]
Breast Cancer	Amplification of HER-2	[21]

EBV infection. EBV-encoded small RNAs (EBERs) are expressed at very high copy numbers in latent infection and can now be detected in a fully-automated setting in archival FFPE material (Figure 6.3b).

ICC remains the most useful technique for studying the phenotypic expression of a particular cell population, due to the availability of an extensive range of antibodies, detection systems and automated equipment. In the analysis of secreted proteins, such as cytokines/chemokines, ICC is of limited use in determining the cell of origin, but RNA ISH has been used successfully to study the spatial expression patterns of such RNA transcripts [28]. There are other examples where RNA ISH can be employed in this manner, and indeed as a validation tool for confirming the results of an ICC experiment RNA ISH has proved very useful (Figure 6.4).

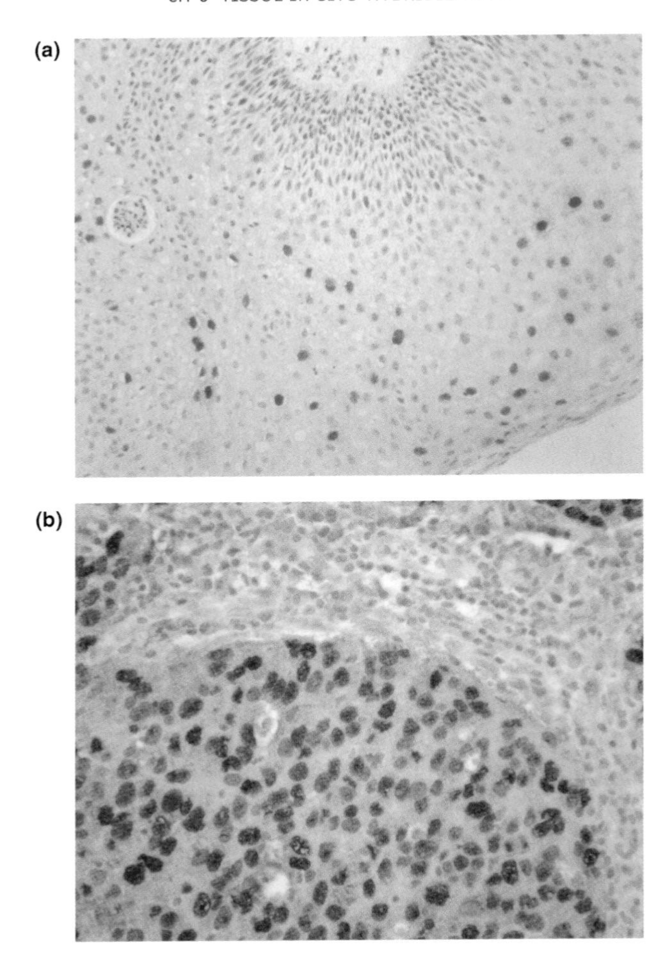

Figure 6.3 The application of ISH to the detection of viral infections; (a) DNA CISH using a biotinylated HPV6/11 probe showing HPV 6/11 infection (red) in a squamous papilloma from the larynx; (b) RNA CISH using a fluorescein-conjugated EBER probe demonstrating latent EBV infection (brown) in a nasopharyngeal carcinoma

Another useful application of RNA ISH is in the identification of cells containing uncharacterized gene transcripts. The elucidation of the human genome has provided information on thousands of mRNAs, of which the functions of many are unknown. The speed of gene-specific riboprobe generation – in comparison to the months it might take to produce a sensitive and specific antibody – gives RNA ISH the edge when it comes to providing clues to the functions of these genes. In addition, RNA ISH is useful in confirming the results obtained by differential gene expression techniques [29].

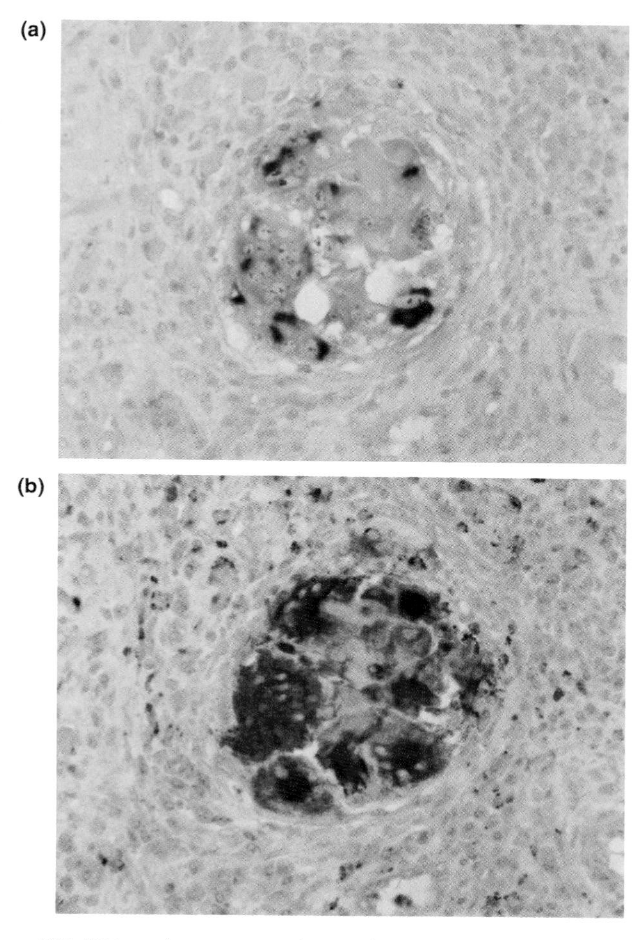

Figure 6.4 mRNA ISH used to support the results of IHC; (a) mRNA ISH using a DIG-labelled probe to demonstrate matrix metalloproteinase-9 mRNA (blue-black) in a cutaneous squamous cell carcinoma; (b) matrix metalloproteinase-9 protein expression (brown) demonstrated by IHC in a serial section of the same sample

6.9 ISH automation

The ISH experimental procedure in the manual format is technically challenging, extremely labour-intensive and prone to human error. With an ever-increasing workload, diagnostic molecular laboratories are required to move faster and more efficiently to meet demands. By automating some or all of the steps in an ISH procedure, accurate control of critical parameters such as temperature, pipetting volume and incubation times can be achieved, leading to high throughput and analysis of many genes simultaneously. An array of semi- and fully-automated systems, from sample processing to results enumeration, is now available.

Certain steps of the ISH assay can be individually automated with modular pretreatment and denaturation/hybridization systems. The VP 2000 processor (Abbott) performs slide dewaxing and pretreatment steps prior to probe hybridization using preprogrammed protocols. The ThermoBrite system (Abbott) is a slide-heating station that holds up to 12 slides and provides 40 user-programmable settings. The Hybaid OmniSlide *In Situ* Thermal Cycler system (Thermo Scientific) is a modular slide-hybridization and *in situ* PCR system. It is designed to allow the processing of up to 20 slides simultaneously. Slides for ISH can be processed through incubation and stringency washing.

Currently there are only a few instruments that provide full automation of the ISH procedure including section dewaxing, pretreatment, hybridization, post-hybridization, and stringency washes and counterstaining. The Xmatrx system (Abbott Laboratories) can process 40 slides at a time and fully automates all FISH assay steps, including the application and removal of coverslips. The BenchMark XT and Discovery XT systems (Ventana) can process IHC and ISH/FISH tests independently or simultaneously. Protocols can be optimized, with flexible options including extended incubation times and varying temperature settings. The Bond system (Leica Microsystems) is a fully-integrated IHC and ISH station. Each station has a 30-slide capacity split into three independent trays for continuous processing. The GenePaint system (TECAN) uses flow-through chamber technology to automate the steps required for performing ISH/FISH or IHC on cells or tissues.

6.10 Image capture and analysis

A number of digital cameras are available for both brightfield and fluorescent microscopy that provide high resolution with multi-function software to optimize real-time image acquisition and subsequent image analysis.

The VIAS system (Ventana) can integrate real-time image capture and quantitation of positive cells in brightfield applications. The system can shift between computer-assisted image analysis and routine microscopy to aid the user in detecting, classifying and counting cells of interest. The *i*VISION Automated Digital Imaging system (Biogenex) is an automated image analysis system that enables users to capture and quantitate images in IHC and ISH applications.

6.11 Recent developments and future directions in tissue ISH

Over the last decade, ISH has emerged as a powerful clinical and research tool for the assessment of DNA and RNA within interphase nuclei in tissue sections. Improved hybridization protocols, along with extensive probe availability resulting from the Human Genome Project, have shifted the focus from a morphology-based diagnostic approach to one which incorporates both morphologic and molecular characteristics. However, new and improved methodologies are constantly being described.

6.11.1 Dual-colour chromogenic ISH

Although dual-probe FISH is gaining in popularity as a clinical diagnostic tool, it has yet to be completely embraced by the pathology community as it is time-consuming and requires specialized equipment and expertise. The enzymatic detection of two separate probe-target hybrids is now a viable alternative to fluorescent dye detection in DNA *in situ* hybridization. The method, known as dual-colour chromogenic *in situ* hybridization (dcCISH), is based on peroxidase- or alkaline phosphatase-labelled reporter antibodies that are detected using a standard immunohistochemical enzymatic reaction [30]. The advantage of CISH over FISH is that it allows the simultaneous examination of tissue morphology and hybrid signals under brightfield microscopy. In addition, CISH-stained slides can be stored at room temperature with minimal loss of signal intensity over time.

The dcCISH technique is a modification of the standard dual-probe FISH technique where hybridization is followed by immunohistochemical detection of the hybrid signals (Figure 6.5). Probes can be labelled with biotin or digoxigenin, or with fluorochromes such as FITC or Texas red. Following hybridization the probe labels are detected with enzyme-conjugated mouse and rabbit antibodies and an enzymatic reaction with appropriate chromogens and substrates leads to the formation of strong permanent colours (usually red and green) that can be visualized using a X40 objective.

Although FISH can be used for the identification of gene gains, losses, translocations and amplifications, CISH is currently best suited for the identification of gene amplification and high-level gene copy-number gains [31]. It is possible to generate FISH/CISH probes within the laboratory for any genomic region of interest but numerous probes suitable for CISH analysis are now commercially available, including gene-specific probes for HER2, EGFR, CCND1 and TOP2A.

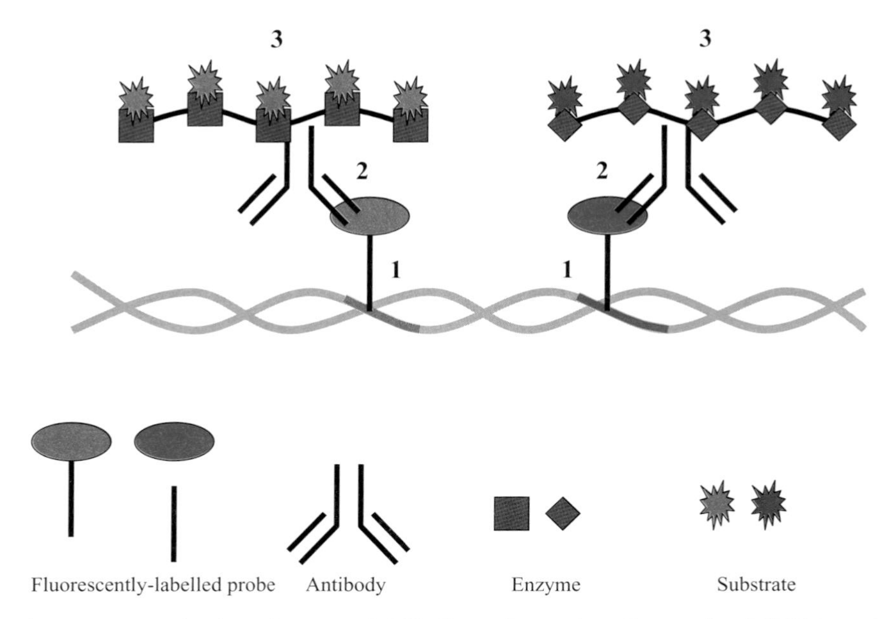

Figure 6.5 Dual-colour chromogenic ISH. Fluorochrome (e.g. Texas red and FITC)-bound probes are hybridized to the target DNA (1). A cocktail of AP-conjugated anti-Texas red and HRP-conjugated anti-FITC antibodies are used to detect the hybrids formed (2). Incubation with an AP-compatible substrate followed by an HRP-compatible substrate (3) provides contrasting colours that can be visualized under brightfield microscopy

6.11.2 Locked nucleic acid for miRNA detection

MicroRNAs (miRNAs) are a class of small (~22 nt) noncoding RNAs that post-transcriptionally regulate mRNA expression of protein-coding genes [32]. There are over 690 human miRNA sequences recorded in the miRBase Registry up to September 2008. There is growing evidence that miRNAs play an important role in the control and function of a number of biological processes such as development, differentiation, cell proliferation, apoptosis, metabolism and proliferation. miRNAs have also been implicated in viral infections, cardiovascular disease and the pathogenesis of cancer. The detection of disease-specific miRNAs could represent the next step in tumour classification and provide a target for possible therapeutic intervention.

The reliable analysis of spatial and temporal miRNA accumulation at the tissue and cell/subcellular levels is essential for the precise understanding of miRNA-mediated processes. The detection of miRNAs by ISH is a relatively new area and by using locked nucleic acid (LNA) nucleotide-containing probes hybridization to short nucleic acids with high specificity can be achieved. LNA oligonucleotides are RNA analogues with high affinity for their complementary DNA or RNA targets [33]. LNA is an RNA-derivative nucleotide analogue in which the ribose sugar ring is fixed in a rigid conformation by a methylene bridge. In FISH analysis, LNA-incorporated oligodeoxynucleotide probes (LNA/DNA probes) have been successfully applied to FFPE and frozen tissue samples [34,35].

6.11.3 FICTION

The FICTION (Fluorescence Immunophenotyping and Interphase Cytogenetics as a tool for the Investigation of Neoplasms) technique was developed in 1992 [36] to allow the simultaneous detection of immunophenotypic markers and genetic aberrations in cell preparations. The original technique was restricted by the number of fluorescent dyes available and the quality of digital imaging. However, improvements in tissue pretreatment methods and the availability of many new fluorescent dyes have seen it used on FFPE material for the analysis of lymphoma [37] and detection of the presence of minimal residual disease [38].

6.11.4 The Allen Brain Atlas

The Allen Brain Atlas (ABA) is a genomic-scale ISH project that has generated a cellular-level gene-expression profile of the adult C57BL/6J mouse brain and spinal cord [39]. This project has used high-throughput CISH with DIG-labelled riboprobes and tyramide amplification to detect mRNA transcripts in tissue sections and map them to the different regions of the mouse CNS. Automated image capture and analysis have provided quantitative expression data that can be directly compared with available microarray data sets. The annotated results for over 2000 genes from this project are freely available. Other similar projects are underway and this approach to genome-wide transcriptional analysis using ISH and correlation with other genome-scale expression-profiling platforms will provide valuable insights into the organization and function of normal and abnormal cells and tissues.

6.12 Conclusions

It is almost 40 years since the first description of the ISH technique. It is now the assay of choice for localization of specific nucleic acid sequences in tissue sections. Over the years, various methodologies and a number of technical advances to optimize the detection of DNA and RNA have enhanced the application and effectiveness of ISH. These include the introduction of high-sensitivity detection systems, improved target-exposure methods, refinement of protocols for co-localization of several targets in the same tissue section, and automated data collection and analysis. Tissue ISH is playing a more and more important role in unravelling the complexity of cancer genetics. With the projected increasing availability of targeted therapies for tumours harbouring specific genetic alterations it is envisaged that tissue ISH will be one of the key molecular assays for accurate patient selection. What remains to be seen is how the use of this technology in the molecular analysis of tissue samples will impact how we detect, diagnose and alter the course of genetic pathology.

Acknowledgements

We wish to thank Ms Josephine Heffernan, Department of Neuropathology, Beaumont Hospital, Dublin, Ireland for the use of Figure 6.1b.

Useful Web sites

miRBase Registry: http://microrna.sanger.ac.uk/sequences/index.shtml

Allen Brain Atlas: http://www.brain-map.org.

References

[1] Schildkraut CL, Marmur J, Doty P. The formation of hybrid DNA molecules and their use in studies of DNA homologies. J Mol Biol. 1961;3:595–617.
[2] Gall JG, Pardue ML. Formation and detection of RNA-DNA hybrid molecules in cytological preparations. Proc Natl Acad Sci USA. 1969;63(2):378–83.
[3] John HA, Birnstiel ML, Jones KW. RNA-DNA hybrids at the cytological level. Nature. 1969;223(5206):582–7.
[4] Langer PR, Waldrop AA, Ward DC. Enzymatic synthesis of biotin-labeled polynucleotides: novel nucleic acid affinity probes. Proc Natl Acad Sci USA. 1981;78(11):6633–7.
[5] Lawrence JB, Singer RH. Quantitative analysis of *in situ* hybridization methods for the detection of actin gene expression. Nucleic Acids Res. 1985;13(5):1777–99.
[6] Logel J, Dill D, Leonard S. Synthesis of cRNA probes from PCR-generated DNA. Biotechniques. 1992;13(4):604–10.

[7] Cox KH, DeLeon DV, Angerer LM, Angerer RC. Detection of mRNAs in sea urchin embryos by *in situ* hybridization using asymmetric RNA probes. Dev Biol. 1984;101(2):485–502.

[8] Stahl WL, Eakin TJ, Baskin DG. Selection of oligonucleotide probes for detection of mRNA isoforms. J Histochem Cytochem. 1993;41(12):1735–40.

[9] Pringle JH, Ruprai AK, Primrose L, Keyte J, Potter L, Close P, Lauder I. *In situ* hybridization of immunoglobulin light chain mRNA in paraffin sections using biotinylated or hapten-labelled oligonucleotide probes. J Pathol. 1990;162 (3):197–207.

[10] Nilsson M, Malmgren H, Samiotaki M, Kwiatkowski M, Chowdhary BP, Landegren U. Padlock probes: circularizing oligonucleotides for localized DNA detection. Science. 1994;265(5181):2085–8.

[11] Pellestor F, Paulasova P. The peptide nucleic acids(PNAs), powerful tools for molecular genetics and cytogenetics. Eur J Hum Genet. 2004;12(9): 694–700.

[12] Heiles HB, Genersch E, Kessler C, Neumann R, Eggers HJ. *In situ* hybridization with digoxigenin-labeled DNA of human papillomaviruses(HPV 16/18) in HeLa and SiHa cells. Biotechniques. 1988;6(10):978–81.

[13] Komminoth P, Merk FB, Leav I, Wolfe HJ, Roth J. Comparison of 35S- and digoxigenin-labeled RNA and oligonucleotide probes for *in situ* hybridization: expression of mRNA of the seminal vesicle secretion protein II and androgen receptor genes in the rat prostate. Histochemistry. 1992;98(4):217–228.

[14] Pinkel D, Straume T, Gray JW. Cytogenetic analysis using quantitative, high-sensitivity, fluorescence hybridization. Proc Natl Acad Sci USA. 1986;83 (9):2934–8.

[15] Smith AM, Dave S, Nie S, True L, Gao X. Multicolor quantum dots for molecular diagnostics of cancer. Expert Rev Mol Diagn. 2006;6(2):231–44.

[16] Srinivasan M, Sedmak D, Jewell S. Effect of fixatives and tissue processing on the content and integrity of nucleic acids. Am J Pathol. 2002;161(6):1961–71.

[17] Bobrow MN, Harris TD, Shaughnessy KJ, Litt GJ. Catalyzed reporter deposition, a novel method of signal amplification. Application to immunoassays. J Immunol Methods. 1989;125(1–2):279–85.

[18] Ness JM, Akhtar RS, Latham CB, Roth KA. Combined tyramide signal amplification and quantum dots for sensitive and photostable immunofluorescence detection. J Histochem Cytochem. 2003;51(8):981–7.

[19] Hartmann C, Mueller W, Lass U, Kamel-Reid S, von Deimling A. Molecular genetic analysis of oligodendroglial tumors. J Neuropathol Exp Neurol. 2005;64 (1):10–14.

[20] Perry A, Fuller CE, Banerjee R, Brat DJ, Scheithauer BW, Ancillary FISH analysis for 1p and 19q status: preliminary observations in 287 gliomas and oligodendroglioma mimics. Front Biosci. 2003;8:1–9.

[21] Pauletti G, Godolphin W, Press MF, Slamon DJ. Detection and quantitation of HER-2/neu gene amplification in human breast cancer archival material using fluorescence *in situ* hybridization. Oncogene. 1996;13(1):63–72.

[22] Turc-Carel C, Aurias A, Mugneret F, Lizard S, Sidaner I, Volk C *et al.* Chromosomes in Ewing's sarcoma: I: an evaluation of 85 cases of remarkable consistency of t(11;22)(q24;q12). Cancer Genet Cytogenet. 1988;32(2):229–38.

[23] Sokolova IA, Halling KC, Jenkins RB, Burkhardt HM, Meyer RG, Seelig SA, King W. The development of a multitarget, multicolor fluorescence *in situ* hybridization assay for the detection of urothelial carcinoma in urine. J Mol Diagn. 2000; 2(3):116–23.

[24] Halling KC, Kipp BR. Fluorescence *in situ* hybridization in diagnostic cytology. Hum Pathol. 2007;38(8):1137–44.

[25] Van den Berg FM, Zijlmans H, Langenberg W, Rauws E, Schipper M. Detection of *Campylobacter pylori* in stomach tissue by DNA *in situ* hybridization. J Clin Pathol. 1989;42(9):995–1000.

[26] Saglie R, Cheng L, Sadighi R. Detection of *Mycoplasma pneumoniae*-DNA within diseased gingiva by *in situ* hybridization using a biotin-labeled probe. J Periodontol. 1988;59(2):121–3.

[27] Stoler MH, Rhodes CR, Whitbeck A, Wolinsky SM, Chow LT, Broker TR. Human papillomavirus type 16 and 18 gene expression in cervical neoplasias. Hum Pathol. 1992;23(2):117–28.

[28] Herbst H, Foss HD, Samol J, Araujo I, Klotzbach H, Krause H *et al.* Frequent expression of interleukin-10 by Epstein-Barr virus-harboring tumor cells of Hodgkin's disease. Blood. 1996;87(7):2918–29.

[29] St Croix B, Rago C, Velculescu V, Traverso G, Romans KE, Montgomery E *et al.* Genes expressed in human tumor endothelium. Science. 2000;289(5482):1197–202.

[30] Laakso M, Tanner M, Isola J. Dual-colour chromogenic *in situ* hybridization for testing of HER-2 oncogene amplification in archival breast tumours. J Pathol. 2006;210(1):3–9.

[31] Lambros MB, Natrajan R, Reis-Filho JS. Chromogenic and fluorescent *in situ* hybridization in breast cancer. Hum Pathol. 2007;38(8):1105–22.

[32] Kloosterman WP, Plasterk RH. The diverse functions of microRNAs in animal development and disease. Dev Cell. 2006;11(4):441–50.

[33] Koshkin AA, Wengel J. Synthesis of novel 2′,3′-linked bicyclic thymine ribonucleosides. J Org Chem. 1998;63(8):2778–81.

[34] Nelson PT, Baldwin DA, Kloosterman WP, Kauppinen S, Plasterk RH, Mourelatos Z. RAKE and LNA-ISH reveal microRNA expression and localization in archival human brain. RNA. 2006;12(2):187–91.

[35] Silahtaroglu AN, Nolting D, Dyrskjot L, Berezikov E, Moller M, Tommerup N, Kauppinen S. Detection of microRNAs in frozen tissue sections by fluorescence *in situ* hybridization using locked nucleic acid probes and tyramide signal amplification. Nat Protoc. 2007;2(10):2520–8.

[36] Weber-Matthiesen K, Winkemann M, Muller-Hermelink A, Schlegelberger B, Grote W. Simultaneous fluorescence immunophenotyping and interphase cytogenetics: a contribution to the characterization of tumor cells. J Histochem Cytochem. 1992;40(2):171–5.

[37] Martinez-Ramirez A, Cigudosa JC, Maestre L, Rodriguez-Perales S, Haralambieva E, Benitez J, Roncador G. Simultaneous detection of the immunophenotypic markers and genetic aberrations on routinely processed paraffin sections of lymphoma samples by means of the FICTION technique. Leukemia. 2003;18 (2):348–53.

[38] Korac P, Jones M, Dominis M, Kusec R, Mason DY, Banham AH, Ventura RA. Application of the FICTION technique for the simultaneous detection of

immunophenotype and chromosomal abnormalities in routinely fixed, paraffin wax embedded bone marrow trephines. J Clin Pathol. 2005;58(12):1336–8.

[39] Lein ES, Hawrylycz MJ, Ao N, Ayres M, Bensinger A, Bernard A *et al.* Genome-wide atlas of gene expression in the adult mouse brain. Nature. 2007;445 (7124):168–76.

[40] Ventura RA, Martin-Subero JI, Jones M, McParland J, Gesk S, Mason DY, Siebert R. FISH analysis for the detection of lymphoma-associated chromosomal abnormalities in routine paraffin-embedded tissue. J Mol Diagn. 2006;8(2):141–51.

[41] Lazar A, Abruzzo LV, Pollock RE, Lee S, Czerniak B. Molecular diagnosis of sarcomas: chromosomal translocations in sarcomas. Arch Pathol Lab Med. 2006;130(8):1199–207.

[42] Fuller CE, Perry A. Molecular diagnostics in central nervous system tumors. Adv Anat Pathol. 2005;12(4):180–94.

[43] Montironi R, Mazzucchelli R, Scarpelli M. Molecular techniques and prostate cancer diagnostic. Eur Urol. 2003;44(4):390–400.

Index

Advanced Techniques in Diagnostic Cellular Pathology Edited by Mary Hannon–Fletcher and Perry Maxwell